Exposure of the American People to

IODINE-131

from Nevada Nuclear-Bomb Tests

Review of the National Cancer Institute Report
and Public Health Implications

Committee on Thyroid Screening Related to I-131 Exposure
Board on Health Care Services
INSTITUTE OF MEDICINE

and

Committee on Exposure of the American People to I-131 from the
Nevada Atomic Bomb Tests
Board on Radiation Effects Research
Commission on Life Sciences
NATIONAL RESEARCH COUNCIL

NATIONAL ACADEMY PRESS
Washington, D.C.   1999

NATIONAL ACADEMY PRESS • 2101 Constitution Ave., N.W. • Washington, D.C. 20418

NOTICE: The project that is the subject of this report was approved by the Governing Board of the National Research Council, whose members are drawn from the councils of the National Academy of Sciences, the National Academy of Engineering, and the Institute of Medicine. The members of the committee responsible for the report were chosen for their special competences and with regard for appropriate balance.

The Institute of Medicine and the National Research Council act under the 1863 congressional charter for the National Academy Sciences to advise the federal government. They also act on their own initiative to identify important scientific and health issues. The president of the Institute of Medicine is Dr. Kenneth Shine. The president of the National Research Council is Dr. Bruce Alberts.

Support for this project was provided by the U.S. Department of Health and Human Services (Contract Number NO1-OD-4-2139, TO 35). The views presented are those of the National Research Council and Institute of Medicine committees and are not necessarily those of the funding organization.

INSTITUTE OF MEDICINE
COMMITTEE ON THYROID CANCER SCREENING

Robert S. Lawrence, *Chair*, Professor of Health Policy and Associate Dean for Professional Education, Johns Hopkins University

Catherine Borbas, Executive Director, Healthcare Education and Research Foundation, Inc., Minneapolis

J. William Charboneau, Professor of Radiology, Mayo Clinic and Mayo Medical School

Virginia A. Li Volsi, Vice Chair for Anatomic Pathology, Hospital of the University of Pennsylvania

Ernest L. Mazzaferri, Professor and Chair, Department of Internal Medicine, Ohio State University

Stephen G. Pauker, Vice Chair for Clinical Affairs, Department of Medicine, New England Medical Center

Henry D. Royal, Associate Director, Division of Nuclear Medicine, Mallinckrodt Institute of Radiology, Washington University School of Medicine

Samuel A. Wells, Professor of Surgery, Washington University School of Medicine

Steven H. Woolf, Private Practice, Family Medicine and Professor, Department of Family Practice, Virginia Commonwealth University

Staff

Marilyn J. Field, Co-Study Director and Deputy Director, Health Care Services, IOM

Cecilia Rossiter, Administrative Assistant, IOM

Vince Knobel, Project Assistant, IOM

Kay Harris, Financial Associate, IOM

Clyde Behney, Deputy Executive Officer, IOM

COMMITTEE ON EXPOSURE OF THE AMERICAN PEOPLE TO I-131 FROM NEVADA ATOMIC BOMB TESTS

William J. Schull *Chair*, Professor, Human Genetics Center, School of Public Health, University of Texas, Houston

Keith F. Baverstock, Head of the Radiation Protection Division, World Health Organization, Rome, Italy

Stephen A. Benjamin, Professor of Pathology, Radiological Health Sciences, and Environmental Health, Colorado State University

Patricia A.H. Buffler,* Dean and Professor of Epidemiology, University of California, Berkeley

Sharon Dunwoody, Professor of Journalism and Mass Communication, University of Wisconsin

Peter G. Groer, Associate Professor, Department of Nuclear Engineering, University of Tennessee

Robert S. Lawrence,* Professor of Health Policy and Associate Dean for Professional Education, Johns Hopkins University

Carl M. Mansfield, Chair, Department of Radiation Oncology, University of Maryland Medical Systems, Baltimore

James E. Martin, Associate Professor of Radiological Health, University of Michigan

Ernest L. Mazzaferri, Professor and Chair, Department of Internal Medicine, Ohio State University

Kathryn Merriam, Synthesis, Incorporated, Pocatello, ID

Dade W. Moeller,† Dade Moeller & Associates, New Bern, NC

Christopher B. Nelson, Environmental Protection Agency, Washington, DC

Henry D. Royal, Associate Director, Division of Nuclear Medicine, Mallinckrodt Institute of Radiology, Washington University School of Medicine

Richard H. Schultz, Administrator, Department of Health and Welfare, Boise, ID

Daniel O. Stram, Associate Professor of Preventive Medicine, University of Southern California

Robert G. Thomas, Kallispell, MT

---

* IOM member
† NAE member

Consultants

Lynn R. Anspaugh, University of Utah

F. Owen Hoffman, SENES Oak Ridge, Inc., Oak Ridge, Tenn.

Donald E. Myers, University of Arizona

Roy E. Shore, New York University Medical Center

Staff

Steven L. Simon, Co-Study Director

Karen M. Bryant, Project Assistant

Doris E. Taylor, Staff Assistant

Catherine S. Berkley, Administrative Associate

Evan B. Douple, Director, Board on Radiation Effects Research

Paul Gilman, Executive Director, Commission on Life Sciences

David M. Livingston, Dana-Farber Cancer Institute, Boston

Editor

Anne Kelly

Sponsor's Project Officer

Charles E. Land, National Cancer Institute

Preface

When the atomic weapons testing program in the continental United States began in 1951 at the Nevada Test Site, the world was sharply divided between two conflicting political ideologies with global intentions uncertain to the other. Each saw its political adversary as a potential aggressor, and each launched a major nuclear weapons development and testing program. In both instances, national defense was seen as taking precedence over the possible health hazards that these programs might present either to the individuals directly involved in the development or testing or to the public generally.

Governmental decisions related to safety both on and off the Nevada Test Site were undoubtedly influenced by a sense of urgency about national security. One apparent consequence is a history of misleading government statements about the Nevada tests. Government decisions were also shaped by the limited amount of information then available on the potential health hazards of fallout, particularly the later-term effects, such as cancer, that would not be quickly manifested. At the beginning of the test period, the Atomic Energy Commission declared that tests did not present a danger to the public, although some staff expressed concern. Experts were not at that time clearly aware that, in addition to direct radiation exposure and inhalation exposure routes, another exposure route—air-grass-milk—was also important.

Not until 1961, near the end of the period of weapons testing did the Federal Radiation Council set as a goal that the annual limit of I-131 doses to the thyroid for a population group not exceed 0.5 rem and that the individual annual limit not exceed 1.5 rem. Although an annual limit of 15 rem had been recommended for members of the public in 1953 (reduced in 1957 to 1.5 rem for minors), the recommendation was not at the time viewed as applicable to fallout from weapons testing.

These perceptions of urgency and safety notwithstanding, in retrospect, it is clear that the exposure of the public was inadequately monitored. This has resulted in large uncertainties in the doses the public may have received and has prompted a continuing and often acrimonious debate about the hazards involved and about the release of information about the weapons tests and their consequences.

The report released by the National Cancer Institute in October 1997 was an effort to assess the hazards of radioactive iodine (more specifically iodine-131, also written as I-131) from the Nevada weapons tests. The analytic challenges the NCI faced in developing its report were formidable and, of necessity, time consuming to meet. Evaluating and quantifying thyroid cancer attributable to Nevada test fallout is an exceptionally complex task that is beset with immense uncertainties related to both the estimates of I-131 exposure and estimates of I-131 related thyroid cancers. Nonetheless, the credibility of the federal government in matters relating to exposure to ionizing radiation may be have been compromised by the agency's perceived slowness in releasing the report, once the analysis was largely completed.

It appears likely that a relatively small proportion of the 160,000,000 Americans alive during the weapons testing program received cumulative doses of iodine-131 greater than the exposure limit set forth in 1961, which would have amounted to 5 rem exposure limit over a ten-year period. However, the number of persons receiving much higher doses, up to 100 rem or more, could have numbered in the tens of thousands. It is the estimation of risk to exposed Americans and the steps that might be taken in response to the risk that has been of primary concern to the review panel established by the Institute of Medicine and the National Research Council.

The panel's work focused on (1) assessing the soundness of the NCI analyses and estimates including those developed separately from the main report, (2) the risk of thyroid disease from iodine-131 fallout, (3) evaluating the benefits and harms of recommending a program of routine screening for thyroid cancer, and (4) identifying strategies for communicating with the public about risks and responses. It reaches a somewhat unsettling combination of conclusions, first, that some people (who cannot be easily identified) were likely exposed to sufficient iodine-131 to raise their risk of thyroid cancer and, second, that there is no evidence that programs to screen for thyroid cancer are beneficial in detecting disease at a stage that would allow more effective treatment. To serve the public interest, the major contribution that the government can make is not to launch an ineffective but politically appealing screening program but rather, to develop effective ways to communicate with the public about iodine-131 exposure and health risks and to involve the public in determining what communication strategies people will find understandable, useful, and trustworthy.

William J. Schull, Ph.D.   Robert S. Lawrence, M.D.
Co-Chair   *Co-Chair*

Acknowledgments

In developing this report, the IOM and NRC committees and staff benefited from the assistance of many individuals and organizations This report has been reviewed in draft form by individuals chosen for their diverse perspectives and technical expertise, in accordance with procedures approved by the National Research Council's Report Review Committee. The purpose of this independent review is to provide candid and critical comments that will assist the institution in making its published report as sound as possible and to ensure that the report meets institutional standards for objectivity, evidence, and responsiveness to the study charge. The review comments and draft manuscript remain confidential to protect the integrity of the deliberative process. We wish to thank the following individuals for their participation in the review of this report: Herbert Abrams (Stanford University), Evan J. Englund (Environmental Protection Agency, Las Vegas), Kristine M. Gebbie (Columbia University), H. Jack Geiger (City University of New York Medical School), Prabodh Gupta (Hospital of the University of Pennsylvania), David G. Hoel (Medical University of South Carolina), Bernd Kahn (Georgia Institute of Technology), Russell Brown (Idaho Falls, ID), Richard A. Kerber (University of Utah School of Medicine), Ralph Lapp (Lapp Inc.), Laura Leonard (Handford Health Information Network), Ross L. Prentice (Fred Hutchinson Cancer Center), Morton Rabinowitz (Media, PA), Marvin L. Rallison (University of Utah), David Ransohoff (University of North Carolina at Chapel Hill), Arthur B. Schneider (University of Illinois at Chicago), Lisa Schwartz (Dartmouth Medical School), Harold C. Sox Jr. (Dartmouth-Hitchcock Medical Center), Lee Wilkins (Columbia, MO), and Stephen Woloshin (Dartmouth Medical School).

While the individuals listed above have provided constructive comments and suggestions, it must be emphasized that responsibility for the final content of this report rests entirely with the authoring committee and the institution. The NRC committee held several public meetings at which presentations were made and public comments were invited. Appendix A lists those who participated. In addition, the committee appreciates the assistance of Lynn R. Anspaugh (University of Utah), F. Owen Hoffman (Senes, Oak Ridge Inc.), Roy Shore (New York University Medical Center), and Donald E. Myers (University of Arizona) who served as consultants.

The IOM committee particularly benefited from the experience and expertise of the presenters and other participants in the workshop the committee convened March 17 and 18 in Washington, DC. Appendix A lists the workshop participants, presenters, and agenda. It also benefited from the timely and thorough review of the scientific literature and other consultation provided by Mark Helfand and Karen Eden. In addition, the committee benefited from the comments of Christopher Merritt, Leslie De Groot, Kenneth Suen, E. Chester Ridgway, and Martin Surks.

Contents

Executive Summary

For nearly 50 years, public concern and scientific debate have surrounded the program of above-ground nuclear weapons testing that the United States conducted in Nevada during the period 1951 to 1962. In August 1997, as the National Cancer Institute (NCI) was preparing to a release a report on exposure to iodine-131 from test fallout, the Secretary of the Department of Health and Human Services (DHHS) asked the Institute of Medicine (IOM) and the National Research Council (NRC) to undertake an independent assessment of the public health and medical implications of the estimated iodine-131 doses received by the American people from this testing and to advise the Department on steps that might be taken in response.

This report presents the conclusions of two committees appointed by the IOM and NRC, which are referred to here collectively as the committee. In developing these conclusions, the committee

- Assessed the soundness of the NCI analyses, which were presented in October 1997 in the two-volume report *Estimated Exposures and Thyroid Doses Received by the American People from Iodine-131 in Fallout Following Nevada Atmospheric Nuclear Bomb Tests.*
- Analyzed evidence linking I-131 exposure to thyroid cancer and other thyroid conditions.
- Reviewed estimates of thyroid cancer cases in the United States that may have resulted from the weapons tests.
- Examined possible clinical and public health responses to I-131 exposure that would be consistent with scientific evidence about the possible benefits and

1

harms of routine screening for thyroid cancer and with research on communicating risk information to the public and the medical community.

• Considered directions for further research.

The committee reached the following general conclusions in three subject areas.

Estimates of national, county, and individual exposure to iodine-131

1. In attempting to fulfill its Congressionally mandated tasks, the NCI undertook a very difficult task that depended on limited data of uncertain reliability and validity. The NCI report represents a careful, detailed, and responsible effort to estimate iodine-131 exposure from the Nevada atomic weapons tests. The NCI's methods were generally reasonable, although specific elements can be questioned (Chapter 2).

2. The estimate of the American people's collective dose from I-131 is consistent with the committee's analysis and is unlikely to greatly over- or understate the actual levels (Chapter 2 on Validation and Uncertainty in Collective or Average Dose Estimates).

3. The levels of detail presented in the report, specifically, county-specific estimates of iodine-131 thyroid doses, are probably too uncertain to be used in estimating individual exposure. For the most part, direct measures of fallout for any particular weapons test were made for only about 100 places nationwide (except near the Nevada Test Site itself). Estimates of county-specific exposures may also have little relevance to specific individuals for whom exposure depends on such critical factors as varying individual consumption of milk and other foods and variations in the source of those foods (Chapter 2).

4. Individual-specific estimates of past exposure to iodine-131 from the Nevada tests are possible but uncertain, often highly so, because critical data are often not available or of questionable reliability. A small minority of the population—those who were young children at the time of testing and who routinely drank milk from backyard cows or, especially, goats—had a significant exposure to I-131 (Chapter 2 on From Fresh Milk to Human Intake and Chapter 3 on Normal Thyroid Physiology).

Estimates of cancer risk

5. Exposure to I-131 as a by-product of nuclear reactions can cause thyroid cancer as shown conclusively by the 1986 nuclear accident in Chernobyl, which resulted in high level exposure for many people. The NCI dose reconstruction model indicates that the level of exposure to I-131 was sufficient to cause and continue to cause excess cases of thyroid cancer. Because of uncertainty about the doses and the estimates of cancer risk, the number of excess cases of thyroid

cancer is impossible to predict except within a wide range (Chapter 3 on Thyroid Cancer Risk Based on NCI Estimates of I-131 Doses).

6. Epidemiological analyses of past thyroid cancer incidence and mortality rates provide little evidence of widespread increases in thyroid cancer risk related to the pattern of exposure to I-131 described in the NCI report (Chapter 3 on Epidemiologic Analyses Using Cancer Registries). They suggest that any increase in the number of thyroid cancer cases is likely to be in the lower part of the ranges estimated by NCI. The epidemiologic analyses are, however, subject to many limitations and uncertainties.

7. Given the uncertainties in both the dose reconstruction model and the epidemiological analyses, further epidemiologic studies will be necessary to clarify the extent to which Nevada tests increased the incidence of thyroid cancer (Chapter 2). Pending these studies, it is prudent for DHHS to plan its responses as if excess cases of thyroid cancer have occurred.

8. Individual-specific estimates of the probability of developing thyroid cancer from exposure to fallout from the Nevada testing program are uncertain to a greater degree than the dose estimates because of the additional uncertainty, in particular, about the cancer-causing effect of low doses of I-131 (Chapter 3 on Thyroid Cancer Risk Based on NCI Estimates of I-131 Doses).

9. The type of thyroid cancer, papillary carcinoma, usually linked to radiation exposure is uncommon and rarely life threatening. Even among those with exposure to iodine-131, few will develop thyroid problems (Chapter 3 on Incidence of Clinically Manifest and Occult Disease).

Thyroid cancer screening, public education, and research

10. There is no direct evidence that early detection of thyroid cancer through systematic screening (rather than through routine clinical care) improves survival or other health outcomes (Chapter 4 on Benefits and Harms of Screening for Thyroid Cancer).

11. A program of systematic screening for thyroid cancer is not recommended either in the American population generally or among regional populations believed to have been exposed to iodine-131 from the Nevada tests (Chapter 4 on Committee Findings and Recommendations). For concerned patients who consult their physicians about screening for thyroid cancer, the decision about screening should be jointly made following a discussion of thyroid cancer risks following exposure to iodine-131, and possible benefits and harms of screening.

12. Rather than promoting systematic thyroid cancer screening, the Department of Health and Human Services should focus on a program of public information and education about the consequences of the Nevada weapons tests. It should involve members of the public and health professionals in developing and testing information strategies and materials. This program should explain basic facts about the Nevada tests, the link between I-131 and thyroid cancer (including

symptoms), the problems with estimating individual exposure to iodine-131 and risk of thyroid cancer, and the pros and cons of screening (Chapter 4 on Information and Decisionmaking).

13. Further research related to iodine-131 would be useful in several areas including the risk posed by low levels of exposure, possible differences in radiation-related and naturally occurring thyroid cancers, and people's perceptions and understanding of the benefits and harms of screening (Chapter 6).

HISTORICAL CONTEXT

In July 1945, the United States conducted the world's first atomic bomb test in Alamogordo, New Mexico. Soon thereafter, U.S. bombers dropped atomic bombs on Hiroshima and Nagasaki as part of a strategy to end World War II. After that war's end, the Cold War led to new national security anxieties and prompted the government to initiate a program of nuclear-weapons testing in Nevada.

During 1951-1962, the United States conducted nearly 100 above-ground nuclear weapons tests in Nevada; another dozen tests were conducted at depths below ground where some atmospheric release of radioactive material was possible. In addition to U.S. tests, other nations including the United Kingdom, the former Soviet Union, France, China, India, and, recently, Pakistan have conducted nuclear tests (the latter two nations only tested underground). Other nations' tests in the 1950s as well as U.S. testing in the Pacific have complicated efforts to estimate the fallout of radioactive materials from the Nevada tests and the health consequences of that fallout. Some systematic studies of fallout patterns were initiated with the first tests, but studies of health effects did not begin until much later.

Public concern about radioactive fallout from the Nevada testing began to emerge by the mid-1950s. Congressional hearings were held in the late 1950s and in 1963 to consider possible effects of the fallout on Americans. The limited test-ban treaty of 1963 was one reflection of public concern about nuclear weapons tests. Litigation was another consequence. By the late 1970s, hundreds of damage claims had been filed against the U.S. government alleging that illnesses, primarily cancers, resulted from the nuclear tests.

In the 1980s, continued concerns prompted a number of further studies to reevaluate radiation exposures of the population following weapons tests in Nevada. In 1983, Public Law 97-414, section 7(a) directed the Secretary of Health and Human Services to conduct research and develop estimates of the thyroid doses received by the American people from iodine-131 in fallout from the Nevada atmospheric tests. In 1983, NCI established a task group to assist it in a program of technical and scientific work that extended for more than a decade. The NCI report, *Estimated Exposures and Thyroid Doses Received by the American People from Iodine-131 in Fallout Following Nevada Atmospheric Nuclear*

Bomb Tests, was published in October 1997. Publication of the report followed newspaper stories that presented information from the report's summary and noted that the analyses were still not available to the public some six years after the NCI had completed most of its mandated work. The NCI report did not include estimates of health risks of exposures. These were provided in separate analyses.

ASSESSMENT OF THE NCI DOSE RECONSTRUCTION

In fulfilling its Congressional mandate, the NCI faced a very formidable task of methodology development, information collection and generation, and data analysis. Challenges included a paucity of data directly relevant to dose reconstruction for events occurring nearly 50 years ago and disparities between the availability of data originally collected for other purposes and the requirements for current use in dose reconstruction. A contributing factor was that the potential health effects of I-131 were not entirely recognized in the 1950s.

Deposition of iodine-131 was estimated based on measurements taken at fewer than 100 sites in the entire continental United States. Even for these sites, amounts had to be estimated because technology at that time did not allow for a direct measurement of I-131 alone; only gross (total) beta activity was measured.

Overall, the committee concluded that the NCI report reflects an intensive effort to collect or generate the data needed for a complicated series of analyses, although documentation of methods, analyses, or results was insufficient in a few places. The committee concluded that the NCI was unlikely to have grossly over- or underestimated the collective I-131 dose, but it was less confident that the NCI had realistically determined the uncertainty associated with the estimate.

For the U.S. population overall, the NCI report presents a comprehensive rationale for assuming that the Nevada atomic weapons tests during the 1950s resulted in an average cumulative thyroid dose from I-131 fallout of 0.02 Gy (2 rad) for the American population collectively (approximately 160 million people) and an average cumulative dose of 0.1 Gy (10 rad) for a large part of the population that was under age 20 at time of exposure.

At the individual level, estimates of exposures to iodine-131 from the Nevada test are uncertain to varying degrees. Critical data on events that occurred almost 50 years ago are often unavailable or of questionable reliability. No matter how painstaking the analysis, the NCI report cannot compensate for the gap between the information available and the information needed to develop valid and reliable dose estimates for individuals. In particular, even with sophisticated methods for geographic extrapolation beyond specific measurement sites, the direct measurements of fallout are far too sparse to provide precise dose estimates at the county or state level for the entire continental United States. Although the committee understood the motivations and initial analytic thinking that led the NCI to calculate and to present county level estimates of exposure for every test, it concluded that the level of dosimetry detail provided for individual counties is inappropriate given the high level of uncertainty associated with the estimated doses. Rather than distinguish between average dose received in many of the counties, it is more appropriate to limit distinctions to much larger geographic areas and, even then, to distinguish only broad exposure levels.

ASSESSMENT OF THE NCI ESTIMATES OF CANCER RISK

The NCI's two-volume dose reconstruction report did not include estimates of thyroid cancer risk related to its estimates of iodine-131 exposure from the Nevada weapons tests. Instead, a separate, later memorandum provided estimates of lifetime thyroid cancer risk and excess cancer cases associated with that exposure. In addition to reviewing the NCI estimates, assumptions, and methods, the committee also reviewed the literature on cancer associated with iodine-131 exposure and examined epidemiological data on the prevalence of thyroid cancer in the United States. The most important conclusion from the literature review is that a link between thyroid cancer and exposure to iodine-131 from fallout—long considered uncertain—is now considered to be established based on studies following the nuclear reactor accident at Chernobyl in 1986. Because these studies have focused on children and are necessarily limited in followup time, the implications for adults now in middle age are unclear.

According to the NCI's revised estimates, which are not broken down by state or county, exposure to I-131 from the Nevada atmospheric tests will produce between 11,300 and 212,000 excess lifetime cases of thyroid cancer with a point or central estimate of 49,000 cases. The committee considered the NCI approach to developing estimates of excess cancer cases due to iodine-131 exposure generally reasonable, but the committee did raise questions about certain assumptions. In particular, it noted that there is disagreement within the scientific community about the assumption of dose-response linearity, that is, the assumption that even the smallest dose of iodine-131 to the thyroid results in some excess risk of cancer. Most exposure to iodine-131 following the Nevada tests was low-level exposure for which evidence of cancer risk is very limited.

Several epidemiological analyses provide little support for the higher estimates of excess cancers related to iodine-131, although these analyses also have limitations that restrict their ability to detect the effects of I-131 fallout on the population. Given the limitations, the analyses suggest that the excess of cancer cases is far below the highest value in the estimated range provided by NCI and is probably in the lower part of the range. Additional analysis (based on the assumption that the risk of developing cancer is constant over a person's lifetime) suggests that about 45 percent of iodine-131 related thyroid cancers have already appeared. Although the committee cautioned against using the NCI county-level dose estimates as a basis for assessing individual risk of thyroid cancer, it suggested that the chance of a significant exposure is highest for those who were young children at the time and who routinely drank milk from backyard cows and in particular, goats.

IMPLICATIONS FOR CLINICAL PRACTICE AND PUBLIC HEALTH

In developing advice for DHHS about thyroid cancer screening and other clinical and public health issues, including communication with health care practitioners and organizations, the committee relied on the evidence-based approaches to clinical practice that have been developing over more than two decades. These approaches stress systematic processes for (1) analyzing scientific evidence about the benefits and harms of clinical practices, (2) developing guidelines for clinician and patient decisions about specific health problems, and (3) clearly presenting the assumptions, evidence and rationale for practice recommendations.

From the perspective of clinical practice and public health policy, a critical problem with the NCI estimates of I-131 exposure and cancer risk discussed above is that they provide little help in identifying those individuals with a significantly elevated risk of developing thyroid cancer. In addition, thyroid cancer caused by radiation cannot be distinguished from naturally occurring thyroid cancer.

More precise estimates would not, in any case, significantly change the conclusions reached by the committee against routine screening of the population generally or of subgroups possibly exposed to I-131. Fundamental to this recommendation is the lack of evidence that early detection of thyroid cancer through screening (rather than through routine clinical care) improves health outcomes. The committee noted that papillary thyroid cancer, the most common form of naturally occurring thyroid cancer and the form linked to radiation exposures, has a high survival rate, regardless of cause, when detected in routine clinical practice without screening. Ninety percent of those diagnosed with papillary thyroid cancer are alive 30 years later.

Screening has the potential for harm as well as benefit. Routine screening for thyroid cancer by palpation and, especially, by ultrasound will identify many nodules, most of which will not be malignant. This will prompt further testing by FNA biopsy. FNA biopsies will discover a few cancerous nodules, will not find cancer in most nodules, and will yield a significant proportion of indeterminate or unsatisfactory samples (20 to 30 percent or more) that may lead to unnecessary thyroid surgery for many people who do not have thyroid cancer or who have very small cancers that would never progress to cause health problems.

Given popular fears of cancer and concern about radiation, the often modest reach of public information programs, and conflicting recommendations from other groups, clinicians will likely see some patients who express concern about possible exposure to radioactive fallout and who request screening for thyroid cancer. Although the committee recommends against policies that encourage or promote routine screening, it is essential that clinicians respond sensitively and constructively to concerned patients who come to them seeking advice. Such a response will involve listening to the patients concerns; discussing their possible exposure to iodine-131 and other risk factors for thyroid cancer; explaining that thyroid cancer is uncommon even in people with some exposure to I-131 and that the thyroid cancer linked to I-131 exposure is rarely life threatening; describing the process, benefits, and harms of screening and the lack of evidence showing that people are better off with it than without it; checking patient understanding of the information presented; and, then, jointly deciding how to proceed. Although the committee believed the decision about a physical examination including thyroid palpation should be a joint one, it recommended against screening by ultrasound for patients without symptoms.

COMMUNICATING ABOUT IODINE-131 EXPOSURE AND CANCER RISK

The historical and political contexts surrounding the Nevada testing program and its aftermath combined with the technical nature of the analyses of iodine-131 exposure, health risks, and screening strategies will make it difficult to communicate information in ways that will be perceived as equally believable and understandable by all those concerned about the consequences of the testing program. The above-ground nuclear tests were purposive, man-made phenomena that left behind a toxic residue, albeit most of it short-lived. From almost the beginning of the testing program, governments and residents of areas adjacent to the test sites have engaged in an intermittent, often acrimonious debate about the release of accurate information and the possible health effects of fallout. Evidence of contamination in more distant areas and from other sources (e.g., the Hanford Nuclear Reservation) have prompted concerns beyond areas near the test site.

Two legacies of the testing program and its aftermath are the considerable, and sometimes intense, distrust of the government as a source of information and advice, and the fact that a segment of the public is convinced the health impacts of exposure are significant and severe. These circumstances create significant challenges for DHHS in constructing a credible and effective public communication program.

The potential audiences for a DHHS communication and information effort will include individuals and groups who have, over the course of decades, developed strong beliefs about the risk and about the trustworthiness of their government, people with little past interest or concern who will want more information when they learn of the exposure issue from news media or other sources; elected officials with varying degrees of concern about the issue; and clinicians and public health officials with varying levels of knowledge about thyroid conditions and screening options. It will be important for DHHS to (1) recognize the existence of these different audiences, (2) decide the audiences they most need to reach, and then (3) use a growing body of empirical and experiential knowledge about communicating risks to develop information strategies that are sensitive to the concerns and perceptions of these groups. For example, if DHHS wishes to communicate effectively with people living in the regions where the tests were conducted, it will need to bring representatives from these areas into the information planning process at the very beginning.

The building blocks for the information and communication program should include several kinds of materials. Brochures or the equivalent should be developed for clinicians, public health departments, and others to give to concerned patients. In addition, although many people lack access to computers, an updated Internet site for both the lay public and clinicians can be a convenient, in-depth source of information for many others. Electronic and written sources of information should explain basic facts about the Nevada tests, I-131, and thyroid cancer and should provide basic education about screening tests, including explanations of possible benefits and harms.

The committee recognized the particular challenge of communicating its conclusions about screening in ways that respond to understandable public concerns about the likely (albeit small) risk of thyroid cancer linked to exposure to iodine-131 from the Nevada atomic weapons tests. Screening programs—whether or not they are supported by scientific evidence—can be a popular response to risk. Notwithstanding this appeal, the committee recommends that DHHS concentrate on developing a program of information and education for the concerned public and clinicians that builds on the analyses, conclusions, and information approaches described in the committee report and on the experience the Department and others have gained in developing similar informational materials.

FURTHER RESEARCH

The committee suggests that DHHS consider additional research in several areas. These areas include (a) the relative effectiveness of external radiation versus internal radiation in producing thyroid cancer; (b) the relative malignancy of radiation-related versus spontaneous thyroid neoplasms; (c) the role of genetic events in the development of thyroid cancer, in particular, the role of ret/PTC oncogene as it may affect the nature of the dose-response relationship for thyroid cancer; (d) people's perceptions of the benefits and risks of screening for thyroid and other cancers and the factors affecting such perceptions including the way quantitative information is presented; and (e) the effectiveness of existing programs to communicate radiation risks. The committee considered public comments calling for further research on the total radiation exposure resulting from all radionuclides deposited in radioactive fallout from nuclear tests in both the United States and other countries. It concluded that such research was unlikely to benefit public health and would divert resources from other uses of greater probable benefit and this as well as the cost of such research should be clearly understood before a decision is made to undertake it due to public concern.

1

Introduction

BACKGROUND AND CONTEXT

On January 27, 1951, the United States began a program of atmospheric testing of nuclear weapons-related devices in Nevada that continued intermittently until August 5, 1963, when the Limited Test Ban Treaty was signed. For nearly fifty years now, the extent and effects of fallout from those tests have generated much political debate and stimulated numerous scientific investigations. The findings of one extended, Congressionally mandated investigation were reported in October 1997 when the National Cancer Institute (NCI) published the two-volume report *Estimated Exposures and Thyroid Doses Received by the American People from Iodine-131 in Fallout Following Nevada Atmospheric Nuclear Bomb Tests.*

As the NCI was preparing to release its report, the Secretary of the Department of Health and Human Services (DHHS) asked the Institute of Medicine (IOM) and the National Research Council (NRC) to initiate an independent assessment of the public health and medical implications of the iodine-131 (sometimes referred to here as I-131) doses received by the American people and to identify steps that might be taken in response. After consultation, the IOM and the NRC agreed to establish an expert panel to undertake this assessment beginning in October 1997.

This report presents the results of that study. The rest of this chapter offers a brief historical overview, describes the study approach, and provides preliminary technical context for the chapters that follow. To help readers who are not expert in radiation dose reconstruction and related topics, this report includes a glossary of terms. It also includes an extensive set of references to the relevant scientific literature and supplementary appendixes.

HISTORICAL OVERVIEW

During the 1951-1963 period of testing at the Nevada Test Site, nearly 100 nuclear weapons tests were conducted above ground (DOE 1994), and another dozen or so were conducted underground at depths from which some atmospheric release of radioactive material was possible. Well before the Nevada tests began, the United States had the world's first test of a nuclear weapon at Alamogordo, New Mexico in July 1945. Soon thereafter, U.S. bombers dropped atomic bombs on Hiroshima and Nagasaki as part of a strategy to end World War II.

Following the war's end in 1945, the United States began testing nuclear weapons in the Pacific. From 1946 to 1958, 66 nuclear tests were conducted in the Marshall Islands, 12 near Johnson Atoll and 24 near Christmas Island (Simon and Robison 1997). All were atmospheric tests, except for four that were detonated underwater. After the first five Pacific tests, the United States began nuclear testing in Nevada.

Because testing in the Pacific was logistically complex and expensive, the government eventually moved all tests to the Nevada Test Site (NTS), which is located northwest of Las Vegas. This location was selected on the basis of such factors as safety, climate, geology, security, population density, and transportation. A few tests were also conducted at other locations in Nevada, New Mexico, Colorado, Mississippi, and Alaska. In addition to the U.S. testing during the 1950s, the Soviet Union and the United Kingdom also conducted atmospheric and other tests that produced a global fallout of radioactive materials. Subsequently, other nations including France, China, India, and, most recently, Pakistan conducted tests (the latter two involved underground tests only). This later testing has not added greatly to the exposure of the American public, but the additional testing in the 1950s (including a large program of American tests in the Pacific) has complicated efforts to estimate both the fallout of radioactive materials from the Nevada tests and the health consequences of that fallout. In addition to nuclear weapons tests, the backdrop for current concerns about exposure to I-131 includes the 1986 nuclear accident at Chernobyl and its aftermath, which includes many cases of thyroid cancer in exposed children.

Concern about radioactive fallout north and east of the Nevada Test Site began to emerge soon after weapons testing began. During the earliest years—before 1954—surveillance offsite of the NTS was performed by the Los Alamos Scientific Laboratory and the U.S. Army. The U.S. Public Health Service continued the offsite monitoring during the years 1954 through 1970. After 1970, monitoring activities offsite of the NTS became the responsibility of the Environmental Protection Agency under interagency agreements. Monitoring techniques included exposure-rate measurements along the roads in the counties surrounding the NTS, a fallout collection program using gummed-film devices in nearly every state, aerial monitoring by aircraft (begun in 1963), and other types of air and precipitation measurement (EPA 1984). Systematic studies of fallout patterns were conducted from the earliest test. By the late 1950s, national programs to monitor milk and food contamination had been established because it had become clear that exposure to radioactive iodine represented the major health threat from weapons testing fallout. Such exposure affected primarily the thyroid gland and resulted mostly from the consumption of milk contaminated with iodine-131.

Congressional hearings were held in the late 1950s and in 1963 to consider the effects of radioactive fallout on the public. One of the earliest reports to estimate radiation dose to the thyroid from radioactive fallout was issued in 1963 (Knapp 1963). The limited test-ban treaty of 1963 was another reflection of public concern, and litigation was yet another consequence. By the late 1970s, hundreds of damage claims had been filed against the U.S. government alleging that illnesses, primarily cancers, resulted from the nuclear tests.

In the 1980s, continued concerns prompted a number of further studies to reevaluate radiation exposures of the public from fallout. In 1983, Public Law 97-414 directed the Secretary of the Department of Health and Human Services to conduct research and develop estimates of the thyroid doses received by the American people from iodine-131 in fallout from the Nevada atmospheric tests (see Box 1.1). (This legislation addressed other issues unrelated to radiation, one of which gave rise to its common labeling as the Orphan Drug Act.)

Following enactment of this legislation, the Secretary requested that NCI respond to the mandate. In 1983, NCI established a task group to assist it in a program of technical and scientific work that extended for more than a decade. It was quickly obvious to the NCI investigators and their advisory group that individual-specific doses could not be calculated with the information available. The advisory group, however, "suggested that it might be possible to estimate, for each atmospheric nuclear weapons test, the iodine-131 exposures from fallout for representative individuals and for the population of each county in the contiguous U.S." (see NCI [1997a], Volume 1, page 1.1). This became the task undertaken by the NCI investigators. The probability of causation tables referred to in P.L. 97-414 were prepared by the NIH Ad Hoc Working Group to Develop Radioepidemiological Tables and were published in 1985 (NIH 1985).

The October 1997 publication of the NCI report was preceded by the appearance of newspaper articles that presented information from the report's summary. The articles noted that the analyses were still not available to the public nearly 15 years after the 1983 Congressional mandate and several years after the NCI had completed much of its work.

For the approximately 160 million people living in the United States during the 1950s, the NCI report estimates the average cumulative thyroid dose from fallout was about 0.02 Gy (2 rad, see Glossary). However, individual-specific doses can vary substantially and to enable individuals to estimate their dose, the NCI constructed the Web site (http://rex.nci.nih.gov) mentioned earlier in this text. In principle, an individual can consult this site to determine his or her thy-

BOX 1.1
Provisions of Public Law 97-414, Sections 7(a) and (b),
Enacted January 4, 1983, Related to I-131

The law directed the Secretary of Health and Human Services to

1. Conduct scientific research and prepare analyses necessary to develop valid and credible assessments of the risks of thyroid cancer that are associated with thyroid doses of Iodine-131;
2. Conduct scientific research and prepare analyses necessary to develop valid and credible methods to estimate the thyroid doses of Iodine-131 that are received by individuals from nuclear bomb fallout;
3. Conduct scientific research and prepare analyses to develop valid and credible assessments of the exposure to Iodine-131 that the American people received from the Nevada atmospheric nuclear bomb tests;
4. Prepare and transmit to the Congress within one year after the date of enactment of this Act a report with respect to the activities conducted in carrying out paragraphs (1), (2), and (3).

In addition, the law directed the Secretary to

5. Construct radioepidemiological tables that estimate the likelihood that persons who have or have had any of the radiation related cancers and who have received specific doses prior to the onset of such disease developed cancer as a result of these doses. These tables shall show a probability of causation of developing each radiation related cancer associated with receipt of a large range of doses in terms of sex, age at time of exposure, time from exposure to the onset of the cancer in question, and such other categories as the Secretary, after consulting with appropriate scientific experts, determines to be relevant. Each probability of causation shall be calculated and displayed as a single percentage figure.

roid dose (based on age, residential history, and recall of milk consumption) from the Nevada Test Site weapons tests.

On the basis of the average doses calculated by county for different age groups, NCI staff initially estimated that 7,500-75,000 exposed persons might develop fallout-associated thyroid cancers during their lifetime (see Appendix B), but subsequent efforts by Land to take into account uncertainties in these estimates yielded a range of 11,300-212,000 cancers with a central or point estimate of 49,000 excess cases (C. Land, communication to NAS, 12/19/97). NCI staff and others have also undertaken epidemiological analyses to see if they can provide confirmation for these estimates (see Chapter 3 and Appendix D).

ORIGINS OF STUDY

In August 1997, even before the NCI published its report, the Secretary of Health and Human Services asked the Institute of Medicine (IOM) and the National Research Council (NRC) to undertake an independent assessment of the public health and medical implications of the findings presented in the report and to consider steps that might be taken in response. The IOM and the NRC agreed, and work began in October 1997. The study was to

1. Assess the soundness of the radiation-dose reconstruction including the methods and data used by the National Cancer Institute, the estimates of thyroid doses in the population exposed, and the estimates of the number of cancers expected;

2. Provide a preliminary assessment of the public-health and medical implications of the NCI thyroid-dose and cancer-risk estimates, in absolute terms and relative to other known risks, and evaluate the potential for other health effects on the thyroid and other tissues;

3. Provide information that will enable DHHS to educate and inform members of the public, especially those likely to have been most heavily exposed at the most vulnerable ages, as to what the NCI estimates and their attendant uncertainties mean for the individual, what the risks are relative to other environmental risks, and what types of appropriate actions they should take;

4. Develop recommendations for various research strategies (epidemiologic, animal, in vitro, molecular, or others) that could refine the risk estimates and thus reduce the uncertainty of the estimated effect of I-131 exposures on public health;

5. Provide information that will enable DHHS to educate members of the medical profession about the NCI report and the limits of present knowledge regarding the carcinogenic potential of I-131; and

6. Develop recommendations that will help DHHS address the implications identified in objective 2 above in terms of

—intervention, surveillance, education, and information strategies for public-health authorities and health-care providers and

—clinical-practice guidelines for evaluating, treating, and counseling exposed persons.

To undertake the study, the IOM and the NRC established two committees with partially overlapping membership (see rosters at the front of this report). Although the IOM and NRC supported both committees, for convenience this report refers to one committee as the IOM committee and the other as the NRC committee. The IOM committee developed the analyses and guidance on clinical and public health policies and communication presented in Chapter 4. The NRC committee developed the other analyses and guidance and reviewed and approved all the findings of the report. Both committees approved the conclusions and recommendations presented in the Executive Summary. Appendix A describes the committees' activities.

OVERVIEW OF TECHNICAL APPROACH AND REPORT CONTENTS

The tasks facing the NCI in 1983—and, subsequently, the IOM and the NRC—were complex. That complexity is reflected in the chapters of the NCI document as listed in Chapter 2 of this report.

As described in Chapter 2, the first technical task for the NRC committee was to review the methods and findings presented in the NCI report. In capsule form, this involved examination of the validity of the use of data from gummed-film collectors as a basis for estimating fallout deposition and related iodine-131 doses, the thoroughness of the identification of the possible pathways of iodine-131 exposure, the reliability and completeness of the measurements available to assess average exposures at the county level, the credibility of the data on milk production and average consumption, the data and soundness of the model to estimate the transfer of iodine into cows' milk, and the other procedures used in the estimation of the doses to representative individuals.

In addition to reviewing the NCI dose reconstruction, the NRC committee considered the estimates of cancer risk that had been presented in a memorandum separate from the main NCI report (see Appendix B). As discussed in Chapter 3, the task here involved reviewing evidence about the link between radiation exposure and thyroid cancer and other thyroid disease; assessing assumptions used in estimating cancer risk; and considering the epidemiological or experimental support for the estimates of cancer risk and the usefulness of further analysis of epidemiological data such as those available in cancer registries supported by NCI and several states.

Chapter 4 examines the criteria for making recommendations about screening for thyroid cancer and reviews evidence about the potential benefits and harms of screening portions or all of the exposed population for thyroid cancer. In developing advice for DHHS about thyroid screening and other possible clinical and public health steps, including communication with the public and with clinicians, the IOM committee relied on the evidence-based approaches to clinical practice that have been developing over more than a decade. The central elements of these approaches stress systematic processes for analyzing scientific evidence about the benefits and harms of clinical practices and developing guidelines for clinician and patient decisions that make clear the assumptions, evidence, and rationale for practice recommendations. Appendix F provides the literature review that provided much of the basis for the analysis in this chapter.

In Chapter 5, the focus is communicating with the public about the information and conclusions presented in both the NCI and the IOM/NRC report. The key conceptual and technical issues examined include the characteristics of risk communication, the matter of credibility of sources of information, the effectiveness of different strategies of risk communication, the need to involve the affected parties in the communication process, and the need to assess the effectiveness of different ways of formulating risk and communicating it to the various interested publics. An example of a method to assist individuals in estimating their personal thyroid cancer risk is provided in Addendum 5. As historical context relevant to understanding government actions and policies during the period of atmospheric weapons testing, a summary of dose limits for exposures of the thyroid for members of the public during this period is included in Appendix E.

Chapter 6 briefly considers directions for future research that might refine the risk estimates and reduce their inherent uncertainties.

CONCLUSIONS

An important determinant of the public's reaction to the information about iodine-131 exposure and risks associated with that exposure is that the aboveground nuclear tests were purposive, man-made phenomena that left behind a toxic residue. Since the tests ended, governments and residents of areas adjacent to the test sites have engaged in intermittent, often acrimonious debate about the possible health effects. The historical and political contexts surrounding the Nevada testing program and its aftermath combined with the technical nature of the analyses of iodine-131 exposure, health risks, and screening strategies will make it difficult to communicate information in ways that will be perceived as equally credible and understandable by all those concerned about the consequences of the testing program. The rest of this report offers guidance to the Department of Health and Human Services, but it is only one step in the process of developing credible and helpful communication with the public.

2

Review of the NCI Radiation Dose Reconstruction

The National Cancer Institute's report consists of two volumes. One sets forth the history of the study, its methods, and findings, and the second consists of extensive tabulations of the data on which the results and findings are based. The second volume does not include the gummed-film data that are said to be archived at the Health and Safety Laboratory (U.S. Department of Energy Environmental Measurements Laboratory) in New York City.

The purposes of this chapter are to determine

• Whether what NCI was asked to do had a reasonable chance of success, given the sparseness of the data and the difficulties involved in reconstructing events that occurred 40-50 years prior.
• Whether there are identifiable weaknesses in the techniques or assumptions used to perform the dose reconstruction.
• Whether the amount of detail presented in the report about thyroid dose estimates of representative individuals by county, by age group, and by milk-drinking habits is meaningful and useful to the public and the public-health establishment.
• Whether the dose reconstruction produced gives useful information about an individual's thyroid dose and, by implication, his or her personal risk of thyroid cancer.
• In addition, we present a summary of estimated doses from the NCI report to illustrate variation in exposure by geographic location and by year of birth.

In this chapter, the committee first lists the contents of the NCI report and gives general impressions of the thoroughness of the study and of the soundness of the study's results. Then there is a general description of the pathways of iodine-131 (I-131 or ^{131}I) exposure in humans. Most of the chapter is devoted to a review of the NCI report, and specific comments are made about NCI's methods, their appropriateness, and their shortcomings.

Box 2.1 lists chapter contents from Volume 1 of the NCI report and the appendices of Volume 2. Separate from the published report, NCI staff prepared estimates of the number of people who might be expected to develop fallout-related thyroid cancer (see Appendix B of this report). Additional information is also available on the NCI Web site (http://rex.nci.nih.gov).[1]

BASIC ASSESSMENT

Given the challenges posed by the paucity of the data relevant to dose reconstruction, the disparity between the original purpose of the data collection and the requirements for dose reconstruction, the need to derive amounts of I-131 ingested on the basis of measured deposition of gross beta activity, and the problems associated with drawing conclusions about counties and states based on incomplete information, this committee was generally impressed by the clarity and thoroughness of the NCI report and the obvious attention to detail. This committee does note, however, that some of the procedures used in dose reconstruction were not sufficiently detailed (for example, the discussion of kriging) for the committee to judge fully the efficacy of NCI's work and the accuracy of the results. The task confronting the NCI investigators was obviously formidable, requiring the reconstruction of events—some of which occurred almost a half-century ago—for which direct measurements of dose or even environmental contamination were often limited or nonexistent, particularly in areas of the continental United States not immediately adjacent to the Nevada Test Site.

There is an important way in which the presentation of NCI's findings is inadequate. The report gives very little statistical tabulation of dose estimates that is useful from a scientific or epidemiologic standpoint. The important numbers, averaged dose by birth cohort, used in the Land presentation, for example, cannot be easily verified as being consistent with the report. The report, the annexes, and Internet Web pages do not appear to allow for the ready computation of average doses for broad categories of people or for the uncertainties in those average doses. This makes it difficult to apply the findings of the report to compute, for example, doses for birth cohorts of people living in the regions covered by tumor registries in two different states or cities, as would be needed in the design stages of an epidemiologic analysis. The numbers provided in the map captions could be used to tabulate average dose by birth cohort, for example, but it would be laborious and possibly inaccurate to use the maps to do this by region. The report and

---

[1]To find information on the 1997 NCI report, it is necessary to follow on-screen links at the NCI Web site to "What Was New in 1997?"

BOX 2.1
Summary Contents of the 1997 National Cancer Institute Report *Estimated Exposures and Thyroid Doses Received by the American People from Iodine-131 in Fallout Following Nevada Atmospheric Nuclear Bomb Tests*

Volume 1
Executive Summary
Technical Summary
Chapter 1.   Introduction
Chapter 2.   History of the Nevada Test Site and nuclear testing background
Chapter 3.   Deposition of ^{131}I on the ground
Chapter 4.   Transfer of ^{131}I from deposition on the ground to fresh cow's milk
Chapter 5.   Cow's milk production, utilization, distribution and consumption
Chapter 6.   Methods and input data for calculating thyroid doses to people resulting from the ingestion of cow's milk
Chapter 7.   Methods and data for calculating doses to people resulting from exposure routes to man other than the ingestion of fresh cow's milk
Chapter 8.   Estimated thyroid doses resulting from atmospheric bomb test conducted at the Nevada Test Site
Chapter 9.   Estimation of doses to specified individuals
Chapter 10. Model validation and uncertainty analysis
Glossary
List of figures
List of tables
List of appendices

Volume 2
Appendix 1.   Description of the meteorological model used to estimate ^{131}I depositions per unit area of ground in the absence of environmental radiation site
Appendix 2.   Structural characteristics of the close assessment methodology and origin and content of the database
Appendix 3.   Information on pasture practices
Appendix 4.   Estimates of the volumes of milk (10^3 L) produced, available for fluid use, and consumed in each county of the contiguous United States in 1954
Appendix 5.   Information on milk distribution
Appendix 6.   Metabolism and dosimetry of ^{131}I
Appendix 7.   Evaluation of the influence of the physico-chemical form of ^{131}I on the thyroid dose estimates
Appendix 8.   Initial retention by vegetation of ^{131}I in wet deposition of fallout
Appendix 9.   Information on the main computer codes used in the dose assessment

Web pages are geared toward calculations at the county or individual level, and can be used for that purpose. However, the ability to make confident distinctions in the differences of estimates for individual persons or counties is in doubt, given the very sizable between-county uncertainties. This problem could be fixed in future work on the Web pages by allowing for the computation of regional doses, as well.

For most locations the NCI dose analysis begins with the gummed-film measurement data. This is appropriate despite the uncertainties in the film's retention of deposited radionuclides, possible losses due to washoff, and the like. That said, the estimates are most useful for average doses for representative groups of persons; the wide range of individual conditions causes estimates for specific individuals to be highly uncertain. The state with the greatest reported range in per capita exposure is Idaho, with an estimated average thyroid dose of 0.157 gray (Gy, 15.7 rad) in Custer County and 0.017 Gy (1.7 rad) in Bingham County. In addition to the dose estimate, there is an estimate given of the variability of the dose estimates, as the geometric standard deviation (GSD) for each county average dose estimate (3.9 and 1.4, respectively, for the two counties).

As described in the NCI report, GSD can be used to provide approximate 95 percent confidence intervals for county dose estimates by multiplying (to give the high end) and dividing (to give the low end) the dose estimates by the GSD raised to the power 1.96. Thus, in Custer County the range of the likely average dose is (0.011-2.26 Gy), the range in Bingham County is 0.0088-0.0329 Gy, and there is an overlap in the likely range of dose of 0.011-0.033 Gy between these two extreme counties.

One important question left unanswered by the NCI report is, When ranges overlap like this, is there really any statistical difference between the county estimates? The answer depends on the proportion of the GSD due to uncertainties shared among counties (a systematic component) and the proportion of the GSD that varies from one county to another (between-county variation). For example, suppose all of the uncertainty (the entire GSD) in the dose estimates in these two counties is caused by uncertainties in the total amount of I-131 coming from the cumulative tests. In this case, the uncertainties in the county dose estimates reflect systematic variation only, and even though the 95 percent confidence intervals for the county estimates overlap, it is likely that if the dose estimate in Custer County is too high, the dose estimate in Bingham County is too high as well. In this case, there is no chance that the dose in Bingham County is actually higher than it is in Custer County. If, on the other hand, all of the uncertainties in the county dose estimates are due to between-county variation, there is a small chance (about 6.5 percent, based on standard statistical calculations) that the doses in Bingham County are actually higher than are those in Custer County. The NCI report is silent regarding the differentiation between systematic and within-county variations in dose. This distinction is, nevertheless, important for the epidemiologic and public-health implications of the report and for whether the amount of detail offered by the report is meaningful. In its appraisal of the NCI report, this committee has attempted to determine what effect the techniques used may have had on the estimates of uncertainty provided in the report, and on whether between-county or systematic components of uncertainty would predominate.

DOSES AND RISKS FROM OTHER RADIONUCLIDES

Although the NCI report focuses on exposure to I-131, as its mandate required, it warrants noting that about 200 radionuclides are produced and released from atomic weapons tests, and exposure to these radionuclides could pose health risks to organs other than the thyroid. For the very large tests that were conducted by Russia and the United States, but not at the NTS, there was sufficient energy released to carry most of the radioactive debris into the stratosphere where it resided with a half-time of a few years. This was the mechanism that gave rise to "global fallout," and because the debris came down to earth months to years after its creation, nearly all of the short-lived radionuclides had already decayed. Some of the more prominent surviving radionuclides in global fallout are carbon-14, strontium-90, and cesium-137 that are incorporated into environmental media and foodchains and persist for decades. Global fallout is dispersed broadly over the hemisphere in which it was injected. Effective dose commitments from global fallout have been evaluated by UNSCEAR (1993). The total effective dose commitment to individuals in the north temperate zone is 4.4 mSv (0.44 rem) per person, of which 2.6 mSv (0.26 rem) is from carbon-14 and only 0.08 mSv (0.008 rem) from iodine-131. Of more interest within the present context is the fallout from the relatively low-yield shots at the NTS. Much of this debris remained in the troposphere, and according to Beck and colleagues (1990) about 25 percent of this material was deposited within the United States within a few days. In this case there was not time for the many additional short-lived radionuclides to decay before being deposited. The primary manifestation of the deposition of the additional radionuclides was a contribution to external gamma dose during the first month after each event. No organ other than the thyroid would have received a larger dose from internal radiation than it received from external radiation (Ng and others 1990). Organs receiving larger internal doses—but less than that from external exposure—were the lower large intestinal wall, bone surface, upper large intestinal wall, red marrow, and stomach wall. Thus, given the public concern about doses from iodine-131, further understanding of the risks to the U.S. population could be gained by a detailed evaluation of the dose and risks from the other radionuclides released by the nuclear tests at NTS. Given that the work has already been done for iodine-131, it would be relatively easy to adapt the method for the additional radionuclides of interest though this exercise is not viewed as an urgent public health necessity.

PATHWAYS TO EXPOSURE OF HUMANS

Each nuclear weapons test produced a large cloud of radioactive fission and activation products and residual nuclear materials, such as plutonium-239. The cloud also included entrained debris from the weapons' casing and varying amounts of other debris, principally soil and the shot tower material if the test was detonated on a steel tower or in such a manner that the fireball touched the ground. Tests conducted aloft, either tethered to a balloon or dropped from the air, contained substantially less debris, and the nearby fallout pattern for balloon or air shots was much less than it was for tower shots. Although the fallout debris was distributed through many layers of the atmosphere, the major part for tower shots would exist in two layers, one at lower elevations that distributed into the lower level wind flow at the time of the test and the other at much higher elevations. Many of those lower level flows moved east into Utah and northward into Idaho. The large particles entrained in the lower stratum fell out over several hours. The upper level debris contained a large fraction of small particles that settled out over large areas as the material moved eastward. Although the patterns varied somewhat, they followed the path of the major air mass present, forming fairly uniform trajectories that compared well with the high-level weather data. Exposure in some counties in the west was attributable to only a few weapons tests, but many counties received deposition from numerous events.

The NCI report (Section 3.6) estimates that only 25 percent of the activity produced was deposited in the United States. Of the remaining 75 percent, much remained in the troposphere and stratosphere and was carried around the globe. However, that portion in the stratosphere decayed before returning to Earth because of the long residence time relative to the half-life of I-131; that in the stratosphere remained entrained for several weeks and a sizable fraction decayed while still airborne; the remainder fell out nonuniformly over the Northern Hemisphere. Even though rainout produced nonuniform patterns and substantial depositions in some locations, the major proportion of the I-131 was deposited in the oceans.

Most of the I-131 deposited in the United States fell on soil or vegetation other than pasture grasses; much that fell on grass was not consumed by cows, and some decayed during the processing and distribution of the milk.

Iodine-131 attaches readily onto dust and other particles and onto vegetation. It is also readily entrained in precipitation and falls to Earth with rain. Once it reaches ground level, a fraction is deposited on grass, dust particles, or by direct absorption into leafy structures. The time-integrated concentration of I-131 on vegetation (per kilogram dry weight) will be several hundred to many thousands of times the time-integrated concentration in ground level air. The large grazing range of cows further integrates or amplifies their accumulated activity into a substantial intake, and the I-131 is taken up in the cow's thyroid and secreted directly into the milk.

A person who consumes cows' milk will concentrate about 25 percent of the intake into the thyroid, which has a mass of about 2 g in a 1 year-old child and about 20 g in an adult. Any I-131 that reaches the thyroid is likely to decay in the gland because the biologic clearance time (the biologic half-life), which is 80-120 days for adults and likely much shorter for young children and infants, is substantially longer than the radioactive half-life (8 days).

Consequently, I-131 deposited over a large area can be concentrated by the pasture-cow-man food pathway into the very small human thyroid to produce relatively high doses of radiation. Any delay or shortened time in any step of this pathway has an equally significant effect on the eventual dose: Time to deposition on the pasture, storing of hay or delaying of pasturing, and time between milking and consumption all affect the radiation dose actually delivered. These are highly variable factors that depend greatly on individual circumstances and the ruminant from which the milk is derived. For example, goats concentrate more I-131 in milk than cows do, and if goats' milk or cows' milk is consumed from private stock, doses will be higher because the milk will be consumed quickly and will not be diluted by other less contaminated sources during pasteurization and distribution. Therefore, the most important determinant of dose for each person was the amount of milk consumed within a period of a few days after the fallout from each test.

SOURCES OF INDIVIDUAL VARIABILITY IN I-131 DOSE

Determination of the doses to members of the public is subject to the highly variable behavior of individuals and the uncertainties in the calculated pathway levels. For these reasons, the NCI report presents doses as averages for representative groups. This is convenient and somewhat useful, but it begs the issue of the variability of dose for individuals as opposed to county or group averages. Only limited information is given in the report on the variability of individual dose estimates around the group or county averages. It is clear, however, that individual dose estimates must be more uncertain than are county averages. As seen in Chapter 6 of the NCI report, age at exposure is important in determining thyroid dose estimates, as is individual milk consumption. Thus, individual doses will vary considerably about the county averages, especially in children. There are two independent components to this variability. First, there will be variation in average doses within counties because of different exposure conditions at different locations and because of differences in the size of the counties and the size and intensity of a storm causing the deposition. Second, there will be variation between individuals with the same exposure conditions because of differences in consumption.

The first source of variation was shown in Scotland to be at least 1 order of magnitude (a factor of 10 times) for the distribution of the isotopes of cesium after the Chernobyl accident (Sanderson and others 1994). Local differences in precipitation rate and runoff on slopes led to variation in cesium contamination of soils and pastures. The variation in I-131 contamination of grasses could be less than that for cesium because soil-to-leaf transfer, and hence the effect of runoff of rain on slopes, is much less important for I-131 than is direct deposition on leaves. Nevertheless, the Scottish experience indicates that marked differences in contamination on pasture can be expected within an area the size of a county.

Variation in absorbed dose from a given exposure situation attributable to differences in the behavior of individuals can be estimated from measurements made after the Chernobyl accident (Likhtarev and others 1995). Measurements of activity in the thyroid glands of exposed persons indicate that, for children of the same age and living in the same settlement, doses can range over at least 2 orders of magnitude (a factor of 100 times) (Goulko and others 1998). Some of this variation will be a function of measurement error (about 25 percent, according to Likhtarev and others 1995), but mostly it will be ascribable to differences in behavior, mainly in the amount of milk consumed. The combined effect of the two sources of variation will set large uncertainties on individual doses even where reliable estimates of contamination at the county level are available.

The NCI report indicates in two places (pp. 9.3 and 9.6) that individual thyroid doses have uncertainties within a factor of about 5 in either direction. If this assertion can be interpreted as a 95 percent confidence interval for true dose, then it corresponds to an individual dose GSD of 2.27, which seems inconsistent with the range of uncertainties given for the county estimates; the average GSD over all counties is 2.24. The near equivalence of these county and individual numbers might indicate that the county GSD actually incorporates uncertainties in individual doses. If this is true, it is not clearly stated in the report or generally reflected in the way the uncertainty analysis appears to have been performed.

The remainder of this committee's commentary parallels the structure of the NCI report. The estimation of thyroid dose due to the cows' milk pathway was broken down into the following basic steps:

- Estimation of the total amount of I-131 released (Chapter 2).
- Estimation of I-131 deposition (Chapter 3).
- Estimation of concentration of I-131 in fresh cows' milk (Chapter 4).
- Estimation of human intake of I-131 in cows' milk (Chapters 5 and 6).
- Estimation of thyroid doses to people from the ingestion of I-131 in cows' milk (Chapter 6).

ESTIMATING THE RELEASE AND DEPOSITION OF I-131

I-131 Released by Tests

Estimation of the errors inherent in the data on the deposition of I-131 on the ground necessarily begins with the reliability of the estimation of the source term, the I-131 actually released in the course of the various atomic weapons tests. One would then proceed to estimate the deposition of radioactive nuclides from the cloud stemming from each detonation and use the deposition estimates to infer individual doses. However, in the NCI report, formal estimation of the source term is limited to those tests for which no network of measurement stations was in operation. For most of the tests, the point of departure in estimating doses to individuals rests on the measured deposition of radionuclides on gummed-film at a network of stations across the United States.

The reliability of I-131 dose reconstruction is obviously greatly influenced by the accuracy of the estimate of the amount released to environmental pathways. The NCI report has prepared an excellent and thorough compilation of the events at the Nevada Test Site, including relevant configurations (air bursts, tower shots, underground tests) and explosive yields in kilotons (kt). Because the explosive power of a device is determined by the number of fissions, the number of atoms of I-131 formed and hence the activity are easily calculated. The NCI report uses a plutonium fission chain yield for mass 131 of 3.72 percent, from which is calculated an activity of 150,000 curies (Ci) of I-131 per kiloton of yield. This is a bit higher than the value of 140,000 Ci/kt calculated by this committee, which used more recent fission yield data (4.3539 percent) provided by the National Nuclear Data Center (England and Rider 1994). It is, however, an adequate and conservative value for the dose reconstruction source term for atmospheric tests. The estimation of the source terms for underground tests is much more complex because varying degrees of venting might have occurred. In this instance, the amount of I-131 released is almost solely that formed by the direct independent fission yield of I-131 rather than the cumulative mass chain yield; the nonvolatile precursors were most likely contained in the ground. Estimated releases for underground tests are, therefore, much more dependent on the air-monitoring data obtained by sampling aircraft and ground stations immediately after the test. These are believed to be good measurements, and the uncertainty in the reported data has been clearly reported.

The NCI report presents an appropriate discussion of the factors that influenced the amount of I-131 present in fallout clouds that left the Nevada Test Site, in particular the influence of the height of the burst on the amount of entrained debris that caused close-in deposition where few, if any, people were present; increased fallout in the near field offsite; and the amounts of activity entrained into the stratosphere as small particles. These variables tend to reduce the amount of radioactivity reaching the more populated downwind areas. Precipitation, or washout, dramatically influenced the deposition patterns as the fallout clouds passed over the United States, and these considerations were also addressed appropriately.

Findings

The source term for the environmental assessment is known to within approximately 10 percent because it is based on weapon yields, which were carefully measured because the yields were fundamental to each test.

Environmental levels of direct relevance to human exposures are less clear because they were modeled, and even though various components of the models are validated with environmental measurements, it is important to note that the data for each pathway to humans were generally not based on comprehensive environmental measurements made for the period of interest.

Measurement of Deposition and Estimation of I-131

Estimates of the deposition of fallout radioactivity are primarily based on data from the gummed-film monitoring network in existence from 1951 until the end of that decade (Beck and others 1990) and, for locations close to the test site, for hand-held-monitoring data. It is estimated that 85 percent of the total release of I-131 occurred while the gummed-film network was in operation. There were 40-95 monitoring stations in operation during most of the period of aboveground testing. Also, during the testing period, ground level measurements of exposure rate were made by the Public Health Service along roads and towns in southern Nevada and Utah. In the mid-1980s, these data were used to form the Town Data Base by the Offsite Radiation Exposure Review Project (ORERP), a study that reconstructed radiation doses from all radionuclides in the area within several hundred kilometers of the Nevada Test Site (Church and others 1990). For a minority of shots, representing a relatively small portion of the I-131 releases, no or few gummed-film measurements were available. For these tests, deposition outside the ORERP area was primarily estimated by meteorologic monitoring and modeling of the passage of clouds, combined with precipitation records from the days of and immediately following the shots. Available data come largely from the activities of the Health and Safety Laboratory (HASL) of the Atomic Energy Commission's New York Operations Office. The hand-held measurements were made by program-associated technicians, scientists, and engineers and should be trustworthy within the limits of the techniques used. Many of the gummed-film analyses were conducted at HASL, which was probably the best such laboratory in the world at the time.

The gummed-film measurements, however, are not really direct measurements of I-131, but rather of gross (total) beta activity. Moreover, the efficiency of the collectors (fraction of fallout measured relative to total deposited on the earth) depended on several factors. Based on an analysis described in Beck and others (1990), the deposition efficiency was assumed in the NCI report to range from 30 percent to 70 percent, depending on daily rainfall. The collection efficiency of the gummed film probably depended not only on daily rainfall but also on the size and solubility of the particles to which the I-131 was attached. The distance from ground level of the nuclear explosion also would influence particle size distribution. Tower shots, for example, would produce larger and less soluble particles than would balloon tests. For most of the soluble substances, a large part of the activity is found in standing water; thus, the extent to which standing water was or was not lost at the time of collection could have influenced the amount of activity recorded (Hoffman and others 1992).

The estimation of the total release of I-131 as well as the conversion of gross beta activity to I-131 was made possible by tables constructed and published by Hicks of Lawrence Livermore National Laboratory (Hicks 1982; 1981). These data were essential to the task of the NCI and have been used for similar purposes by other dose-reconstruction studies (Stevens and others 1992; Ng and others 1990).

Imputation of Deposition When No Direct Measurements Exist

Only a relatively limited number of gummed-film measurements were available during the period of testing, so interpolation methods were devised to give county level estimates of deposition on a test-by-test basis. The simplest of these, the area-of-influence precipitation-corrected (AIPC) method, was used for estimation when too few positive gummed-film measurements were available for the use of the more elaborate kriging algorithm described below. The AIPC method is very simple: the gummed-film measurements from the nearest stations were used in all other counties after applying a correction based on the difference in precipitation rate in the county compared with that at the location of the gummed-film measurement.

The majority of deposition estimates were interpolated from the gummed-film measurements using a mathematical technique called kriging. The real distinction between interpolating using kriging versus using the AIPC method is that kriging involves fitting parameters in a statistical model for the gummed-film measurements prior to performing the interpolation. Because kriging is the most important method used to obtain deposition estimates, much of this section is devoted to reviewing the NCI report's application of this method.

Overview of Kriging

Various substances are spatially distributed, that is they have unique values at each point in space or at each geographic location. Examples are the amount of minerals in the ground or the amount of fallout contamination deposited on the soil. Various methods can be used to estimate the amount of minerals, fallout, etc. at locations where no measurement data exist. The simplest such technique is the average of the measurement data nearby to the site of interest. More sophisticated estimation techniques are usually termed as interpolative methods. One such technique is kriging, named for D.G. Krige, a South African mining engineer and a pioneer in the application of spatial interpolation. Kriging, developed by Prof. Georges Matheron and his associates at the Institute for Mathematical Morphology in Fontainebleau, France is the process and method of estimating the local value of a spatially distributed quantity while considering the interdependence of the measurements within the region. The mathematics of this technique are designed to minimize the error in estimation by using information about the similarity ("covariance" in mathematical terms) of adjacent data points. Interested readers can consult various introductory texts on this subject (see, for example, Isaaks and Srivastava 1989).

Kriging is considered an appropriate technique for the interpolation of data measured at a small number of locations and where some degree of spatial correlation is exhibited; that is, when values at locations in close proximity are assumed to be more similar than are values at locations far apart. As discussed by Beck and others (1990), the HASL gummed-film stations provide a set of measurements that have both characteristics, and thus, kriging is a reasonable technique for estimating deposition of I-131 from the Nevada Test Site.

Kriging, in any specific setting, involves considerable modeling before interpolations are computed. The answers supplied by kriging will depend in important ways on choices made by the analyst, generally with incomplete knowledge about which choice is correct for a given situation. The modeling issues to be dealt with include:

• The possible selection of a transformation of the measured data, typically used to improve the symmetry or normality of the overall distribution of measurements.

• The use of additional covariates available at the data locations and at the points of interpolation to supplement or improve the interpolations.

• The form of the function or the variogram used to describe the pattern of spatial correlation between measurements.

In practice, the variogram model is a positive linear combination of a small number of standard models that still allows for variation in the parameters of the standard models. The range of dependence is one parameter to be estimated. Having made these modeling choices, parameters in the regression function and variogram were estimated for each day. The resulting kriging estimator or predictor at a specific point is a linear function of the data involving the spatial correlation. In this sense, kriging is similar to multivariate regression analysis, but with the extra complexity involved in modeling spatial correlation.

In most uses of kriging, the kriging estimator is exact; if the data value is known at a location then the kriging estimate at that location will coincide with the data value. This property was used by the authors of the NCI report to validate the variogram model. To estimate the magnitude of the error in the interpolations, the authors sequentially deleted each location from the data set and used only the remaining data locations to estimate the value at the deleted one; the estimated values were then compared with observed values.

Kriging and the NCI Study

The kriging methods used in the NCI report used the following choices: a logarithmic transformation of all the nonzero measurements was made to improve the normality of the distribution of the gummed-film measurements. Adjustments were then made to incorporate information pertaining to precipitation in each county for the day in question. If more than one measurement of rainfall existed for a particular county, the arithmetic average was used, and if no measurements were available, assignment of a precipitation value was based on the closer measurements in the adjacent counties. It appears that residuals were computed and then kriged and added to the precipitation component, but details are not given in the report. A data-driven approach was used in modeling the spatial correlations; variograms were chosen from among several candidates to give the best fit to a single day's gummed-film measurements using an average error criterion (C. Gogolak, private communication).

The value of kriging for estimating I-131 deposition depends, first, on the quality of the measured data—here the gummed-film estimates of deposition at HASL stations—and, second, on the number and location of the sampling stations. Accordingly, the evaluation of the adequacy of the gummed-film data and the number of monitoring stations is essential to an overall appraisal of whether the task set for NCI was likely to succeed.

The principal discussion of the quality of the gummed-film measurements appears in a publication by Beck and others (1990). Table 7 in that report gives data that compare estimates of cumulative cesium-137 derived from gummed-film beta-activity data with direct measurements of cesium from analyses of soil samples. The correlation between the two estimates is extremely high (~0.95). Data are given only for 11 stations, not all of which have complete data, but the comparison certainly adds credence to the deposition estimates based on the gummed-film data. Beck and others (1990) estimate that the errors in deposition estimates for a given site and shot are probably no worse than a factor of 2 and that uncertainties in estimates of cumulative deposition are much lower. Nevertheless, that discussion—along with the data in Table 7 of that report— indicate that, even for total deposition, the fraction of variability attributable to measurement error could still be a significant portion of total deposition variability, perhaps 50 percent, depending on how the rather vague comments about uncertainty are interpreted. Further, Beck and co-workers claim that errors in the gummed-film estimates are essentially random. Kriging can be modified to incorporate random error in the measurements themselves. This was not done by NCI collaborators, although some ad hoc adjustment to the kriging estimates of variability (see Table 3.7, NCI report) were made to account for errors in both the gummed-film measurements and the precipitation index used as a covariate in the analysis. A more systematic incorporation of measurement error into the kriging would have altered the interpolation estimates as well as the variability of the estimates. In this case, the kriging estimator would not be exact at data locations, but theoretically would improve the predictions relative to methods that ignore the measurement error. Incorporation of measurement error into the kriging generally requires outside information concerning the magnitude or variance of measurement error at each data location and can complicate assessments of prediction error.

Turning to the number and location of the HASL stations, the adequacy of the monitoring system for the production, at the county level, of reasonable estimates either of deposition from a specific shot or of total deposition, depends greatly on the basic structure of the spatial correlation among measurements. If spatial correlations are high, then few measurements will be needed to characterize deposition, even in counties far from the locations for which measurements are available. If they are low, then trying to form county-specific estimates is an exercise in futility, even though estimates over much wider areas—states, regions, or the country as a whole—can be reasonably accurate. The report gives no specifics concerning the spatial correlation pattern of the gummed-film data. Beck and others (1990) give gummed-film data for several locations for several important tests (Table 5, Beck and others 1990), so some spatial correlation is observed in the data, although no specific analysis is provided.

Magnitude of Error

Pages 3.29 and 3.30 in the NCI report present some general remarks about the magnitude of error in the estimates of deposition obtained by kriging. The report finds that the predicted values, as estimated by a variety of means, are generally accurate within about a factor of 3 and do not appear to contain any significant bias in either direction. This factor-of-3 estimate is by itself difficult to interpret. Apparently, it refers to the magnitude of error in the estimates of a single day's deposition in a specific county. It is not clear, however, whether the factor of 3 is to be regarded as a 95 percent confidence interval or as a single standard deviation. The committee assumes that it is a 95 percent confidence interval, specifically, that the GSD for a single day's county estimate is equal to 1.75, as this usage is consistent with the rest of the NCI report.

The NCI estimate of the magnitude of error is apparently a single summary statistic based on the magnitude of the prediction error from the cross-validation of the kriging estimates. The approach by which this estimate of uncertainty was obtained seems reasonable. However, throughout the NCI report there is no attempt to apportion the uncertainty estimate (GSD = 1.75) into systematic versus between-county variability, although it is likely that each type of variability exists to some degree. In general, one could expect that adjacent counties will have considerable sharing of uncertainties, so they are not independent. For counties separated by larger distances, errors in the county estimates should be independent. Without further information about the type of variogram models used—the rate of decay of the spatial correlations—it is difficult to determine from the report whether counties within the same state are correlated enough that systematic or between-county estimates, as described above, dominate.

Findings and Recommendations about Kriging

This committee finds that kriging is appropriate for the task of interpolating the gummed-film measurements. There appears to be little question about the general validity of wide-area deposition estimates, such as regional or nationwide total deposition estimates. Estimates made for local areas for which good coverage by the HASL gummed-film stations exists, such as the cities reported in Table 7 of Beck and others (1990), also are regarded as reliable. Nonetheless, it is possible that different analysts using the same general class of techniques would derive different answers by choosing different models. Although it seems unlikely that important differences, such as those in wide-area average deposition estimates, would result, estimates at the small scale could be affected considerably. For example, the explicit incorporation of random error in the gummed-film estimates at the HASL locations might lead to much less variability in the interpolation estimates at the county level.

There is insufficient information given in the report or appendices to determine whether county level deposition estimates are supportable and, in particular, whether variation at the county level in the deposition estimates represents anything but statistical noise.

Recommendation: It would be desirable for NCI to ensure that the data are preserved in electronic form and made available for possible re-analysis.

Recommendation: To the extent possible, more details should be published on the variogram modeling as well as on the actual application of kriging, and the computer codes used in the NCI report should be made available.

FROM I-131 DEPOSITION ON THE GROUND TO COWS' MILK

Figure 3.34 in the NCI report is a map of I-131 deposition estimates for all tests for all counties in the contiguous United States. Figure 4.25 of the report gives the estimates of total time-integrated concentrations in fresh cows' milk, IMC. The primary path by which cows' milk is contaminated with I-131 is ingestion of I-131 from contaminated pasture. Unlike the situation with deposition, few measurements of I-131 (see Knapp 1963; Campbell and others 1959) or total beta decay in cows' milk or in pasture were made during the testing. To offset this paucity of measurements, the approach taken to estimating the concentration of I-131 in cows' milk was to break down the transfer process into discrete steps for which experimental or observed data either exist or could be inferred based on expert knowledge.

For a given county, i, moving from deposition to milk concentrations through the pasture pathway involves the following primary estimation steps, done for each day, j, after deposition from any particular test for 60 days thereafter.

- Estimate the initial concentration (day 0) of I-131 in pasture.
- Estimate the remaining concentration of I-131 in pasturage on day j + t (t = 0, ... , 60).
- Estimate the amount of pasturage eaten per cow on day j + t.
- Estimate the fraction of I-131 ingested with pasturage intake that is secreted into milk.

Initial Concentration

The parameter used to define the initial concentration in pasture is the mass interception factor $F^*(i,j)$, the ratio of the fraction of deposition retained in vegetation to the standing plant biomass (kg/m^2). Multiplying $F^*(i,j)$ by deposition, $DG(i,j)$, on day 0, gives the initial concentration, $Cp(i,j,0)$, of I-131 retained in pasture (nCi/kg dry mass). This factor was treated separately for wet and dry deposition. Estimates of $F^*(i,j)$ for periods of dry deposition were based on a review of experimental work by Chamberlain (1970). The value of $F^*(i,j)$ for dry conditions, F^*dry, used in the report depends on the value of standing crop biomass, Y, and on the distance from the Nevada Test Site. The reason F^*dry is assumed to depend on distance is that distance is treated as a surrogate variable for particle size, which has been found to affect mass interception. The relationship between distance from the test site and particle size is based on a limited amount of data, which are summarized by Simon (1990).

The mass interception fraction for wet deposition was found to depend on the amount of rainfall or standing crop biomass. Empirical values were in an experimental program (Hoffman and others 1992; 1989) set up by the NCI though the NCI dose estimates in the 1997 report used an earlier formula derived by Horton (1919) for that parameter. The distribution of total standing crop biomass needed for calculating F^*dry was assumed to be lognormal with a median of 0.3 kg/m^2, and GSD of 1.8 (see p. 4.4, NCI report), based on data from Baes and Orton (1979).

Concentrations in Pasture on Day j + t

Radioactive decay and environmental removal processes are assumed to reduce the initial amount of radioactivity deposited on pasture grass. The NCI report takes as the combined effect of decay and environmental removal an environmental half-life, λ_e, of 4.5 days. Experiments by Hoffman and others (1989) found no important systematic differences in λ_e by physico-chemical form; that is, particle or soluble I-131, plant growth rate, or the fall of subsequent uncontaminated rain.

Amount Eaten by Dairy Cows

The NCI report notes that pasture practices of the 1950s differed from those of today, with more reliance then on outdoor grazing. Reconstruction of pasturing practice of the 1950s relied primarily on Dairy Herd Improvement Association data for several variables, including average weight of cows, milk yield, and dry-matter intake. Expert opinion was obtained to fix the beginning and end of the pasture season and the fraction of dry-matter intake attributable to pasture. Those data were developed for each of 67 pasture regions (NCI report, Figure 4.10) in the contiguous United States. The database was used to estimate the amount of pasture eaten by cows in each county on each day of interest.

Those data were modified for backyard cows by extending the pasture seasons for each pasture region and assuming that backyard cows were fed entirely on pasture during that extended season. However, this committee finds the definition of a backyard cow confusing and possibly misleading. Careful reading of the NCI report is required to understand that Category 1 milk (see below) includes milk from small farms that is consumed on the farm; this is not, in general, the same as milk from a backyard cow. In the 1950s there were just as many, or possibly many more, people living on small dairy farms as there were people who owned one or two cows. It is important to distinguish more clearly the consumption of fresh milk from cows on small dairy farms from milk that came from true backyard cows. On the dairy farms, families drank fresh milk within hours of collection, but the cows would have had standard pasturage and higher milk yield. In the 1950s in high-milk-producing states and areas (for example, in upstate New York) the relatively small dairy farm was common. As the report stands now, it would be easy for someone growing up on a small farm to estimate his or her dose erroneously because of confusion about what constitutes a backyard cow.

Intake from pasturage is converted into a daily pasture intake equivalent, $PI^*(i,j)$, which is the ratio of the total activity of I-131 in pasturage actually eaten to the time-integrated concentration in pasture at the time of deposition.

Percentage I-131 Ingested with Pasturage Intake Secreted into Milk

The cumulative fraction of I-131 after a single episode of intake that is secreted into cows' milk (on a per-liter basis) is estimated to be equal to 5 percent, with a range of 1-20 percent, and it varies not only between animals but also in the same animal at different times. The NCI report describes the transfer of I-131 from pasture to milk primarily in terms of an intake-to-milk transfer coefficient, f_m, defined as the time-integrated concentration of I-131 activity in milk (nCi d per L) per unit of I-131 activity consumed (nCi). Much of the published data (summarized in Table 4.6, NCI report) give f_m estimates based on dozens of feeding experiments, with each experiment reporting results from a relatively small number of animals. The NCI report (p. 4.21) assumes that f_m for I-131 in Nevada Test Site fallout is distributed as lognormal random variables with a geometric mean of 4×10^{-3} d per L for any county in the contiguous United States for any time of year.

The time-integrated concentration of I-131 in milk resulting from a deposition on day j in county i is computed as

$$IMCp(i,j) = DG(i,j) \times F^*(i,j) \times t_e \times PI^*(i,j) \times f_m$$

Similar calculations are provided for other exposure routes.

Findings

The basic NCI approach to estimating, for each county, the median time-integrated concentration of I-131 in milk after a given deposition is sound. Although few or no relevant measurements of I-131 contamination of milk from the Nevada tests exist, the NCI review of experimental work on f_m and $F^*(i,j)$ is extensive, and much work has been done on the major components of the pasture-cow-milk pathway.

The assumption of a uniform weathering half-life of 10 days in all situations is undoubtedly a crude approximation; some evidence that λ_e is less than this (perhaps 3-4 days) in the spring months is provided by the International Atomic Energy Agency (IAEA 1996). The committee would expect f_m to depend on particle size, with higher milk transfer rates for smaller particles, so that eastern areas distant from the Nevada Test Site might have had relatively more of the I-131 in pasturage transferred into cows' milk. This is supported by the relatively high f_m (0.0061-0.0127 d per L) reported by Weiss and others (1975) and Voillequé and others (1981), working with fallout from far distant atomic weapons tests. The reconstruction of pasture practices of the 1950s is inherently less certain than are the rest of the calculations, the likelihood that they are grossly in error seems low, given the extensive effort that has gone into the process. The assumption that hay, rather than fresh pasturage, is uncontaminated, is crucial to the calculations of cows' intake of I-131. This appears to be partly confirmed by the Chernobyl experience; BIOMOVS (1991) found a factor-of-10 difference between I-131 concentration in milk from cows on and off pasture.

There is a point of confusion regarding the uncertainty analysis described at the end of Chapter 4 of the NCI report. The uncertainty analysis is designed to estimate the variability of the county average concentrations of I-131 in fresh cows' milk, not to give more-individualized estimates of variability. Because it is the variability of the county level estimates that is at issue, the method used seems to exaggerate the importance of uncertainties in the milk transfer coefficients and probably in the mass interception factors as well. This error results from the failure to differentiate between the uncertainties of averages relative to the uncertainties of individual items in the average. The report assumes that milk transfer coefficients are distributed as lognormal variables with a geometric mean, 4×10^{-3} d per L, and GSD, 2.1. This appears to describe reasonably well the data in Table 4.6 of the NCI report, but each experiment reported in this table is based on a small number of animals. However, a given county could have contained hundreds or thousands of dairy cows. The variability of average f_m over these larger herds is sure to be smaller than would be the variability in f_m in a single cow.

Nevertheless, there is uncertainty as well as variability between cows in the average f_m that should be assigned to the animals making up the dairy herds of the 1950s. It is likely that values based on relatively recent data are not appropriate for the 1950s. It is also probable that true backyard cows would have had lower milk production and higher f_m than would commercial cows of that time. A recent survey (Brown and others 1997) of 10 experts' interpretations of the likely average value of f_m for future reactor accidents yielded estimates that range from 0.004 d per L to 0.011 d per L, with a wide range of uncertainty attributed to the estimates. This subjective probability assessment, although intended for reactor accidents rather than for weapons fallout, indicates that the value of f_m used in the NCI report (0.004 d per L) is uncertain to a possibly important degree and that not all of this uncertainty is caused by random variability among averages of small groups of animals.

Similar remarks can be made about the mass interception factors. A single county can contain a variety of pastures, and the average $F*(i,j)$ over these types could be much less variable than would be the mass interception factors obtained in single experiments. Although the NCI report does not explicitly incorporate uncertainty in the fraction of intake from pasture, errors in the fraction of intake from pasture could produce both systematic errors affecting large numbers of counties and between-county errors. In general, however, systematic uncertainties, after proper accounting for averaging in this part of the calculations, are likely to be small based on the extensive experimental data available.

FROM FRESH MILK TO HUMAN INTAKE

Figure 6.2 in the NCI report is a map of estimates of integrated concentrations of I-131 in volume-weighted milk for all tests. Figure TS.3 gives the estimated time-integrated concentrations of I-131 in volume-weighted mixed milk. Going from Figure 4.25, which shows integrated concentrations of I-131 in fresh cows' milk, to Figure 6.2, involves multiplying the estimates of each county's integrated concentrations in fresh cows' milk by the estimates described in Chapter 5 of the NCI report for the fluid-use milk production of each county. The estimates in Figure TS.3 for integrated concentrations in volume-weighted mixed milk are obtained from volume-weighted fresh milk by application of milk transfer functions, VOL(i,j), which describe the transfer of milk between counties.

Milk Production

Chapter 5 of the NCI report describes the modeling of the milk production and distribution system for the 1950s, which was used to develop the VOL(i,j) values. Milk production for fluid use and size of milk surplus for each county, assumed constant over the years of interest, were estimated using U.S. Department of Commerce data, U.S. Department of Agriculture (USDA) data, and census data as follows:

- The number of cows in each county in 1954.
- State statistics on the amount of milk used for non human consumption on farms.
- State statistics on the amount of milk used in manufacture of food products.
- Per capita milk consumption in each state.
- Population in each county.

The county level data were imputed by apportionment of state data.

The distribution of fluid milk from counties with surpluses to counties with deficits was modeled. This was done by classifying fluid milk production in each county into four categories corresponding to four population groups:

- Category 1, those living on farms in the county where the milk is produced.
- Category 2, those living in the county where the milk is produced but not on farms.
- Category 3, those living in a group of neighboring counties within a designated milk region.
- Category 4, those living at greater distances in other milk regions.

The model for the distribution system first moves milk from counties with surpluses to neighboring counties with deficits (Category 2 milk), then to other counties in the milk region with deficits (Category 3 milk), and finally from milk regions with surpluses to neighboring milk regions with deficits (Category 4 milk). Time-integrated concentrations of I-131 in milk coming out of the distribution system for human consumption are calculated based on an assignment of delay time between production and consumption of 1 day for Category 1 milk, 2 days for Category 2, 3 days for Category 3, and 4 days for Category 4. The information on which the estimates of volumes and the direction of milk flow within and between milk regions and the data on delay times between production and consumption are based is described as "qualitative."

Findings

The largest source of uncertainties in the county milk production estimates is most likely in the apportionment of state data on the amount of milk used in manufacture to form county estimates. Errors in such apportionment add to between-county uncertainties in the size of county surpluses or to deficits in fluid milk production but would not add to systematic uncertainties because the state statistics were known.

The errors in I-131 concentration coming from incomplete information in the modeling of milk distribution would add to systematic error only if the delay times between production and consumption assumed for each category of milk are incorrect. Even the addition of 2 extra days between production and consumption in all of the categories would reduce average I-131 concentration levels only by about 15 percent.

The uncertainty analysis provided on pages 6.4-6.6 of the NCI report starts with the uncertainties in the county estimates of time-integrated concentrations of milk. These initial uncertainties, as discussed above, might be too large because they appear to neglect the effect of pooling in averaging out variations in such factors as milk transfer coefficients, f_m, except for Category 1 milk, where it is more reasonable that intake depends on the production of only a few cows.

The uncertainty analysis does not explicitly deal with the effect of errors in the VOL(i,j) terms that define the milk distribution system. These would appear to add significantly to between-county uncertainty but not to overall systematic error. Given that kriging has produced estimates of deposition that are generally accurate only to within a factor of 3, much of the variability of which must be within-county rather than purely systematic, the level of detail used in the reconstruction of the milk distribution system seems unwarranted. Nevertheless it reasonably represents many broad features of milk distribution in the country.

Individual Consumption of Milk

Milk consumption is assumed in the NCI report to vary by age, race, and sex, but not by time, over the period 1950-1965. Detailed descriptions of infant, child, and adult milk consumption are based on several reports (Yang and Nelson 1986; Rupp 1980; Durbin and others 1970; Thompson 1966; PHS 1963b; 1963a). Data on milk consumption patterns at the county level are not available. Differences among regions of the country and in the degree of urbanization (city versus country) in milk consumption are used in the milk consumption model based on statistical data for 1954 (USDA 1955). Considerable data on the distribution of milk consumption (based on PHS 1963b), in addition to average milk consumption, are presented in the report and were used in the calculations.

Findings

Uncertainties in the milk consumption data exist principally because they are based on self-reports of individual or family members' consumption, for limited periods (3-7 days), and for finite (although relatively large) samples of subjects. The committee noted some inconsistencies in the self-reports. For example, USDA (1955) reported per capita milk consumption rates by region, using information collected over 1 week in 1955, that were somewhat higher than other USDA estimates based on total amount of milk for fluid use for that year. The reliance on regional statistics for use at the individual county level adds some error in individual county estimates. The likelihood of systematic error affecting regional estimates of consumption appears small, however, reported milk consumption rates were adjusted to correspond to production. The data used in estimating infants' and children's consumption of milk are necessarily somewhat less certain than are data for overall per capita consumption, but again the likelihood of important systematic errors in these data seems small.

FROM INTAKE TO THYROID DOSE

National Cancer Institute county estimates of thyroid dose resulting from consumption of cows' milk are developed according to age and sex, which influence both the amount of milk consumed and the dose conversion factor (mrad/nCi) by which a given time-integrated concentration of I-131 in milk is converted to thyroid dose.[2] Determination of a dose conversion factor involves analysis of the fraction of ingested I-131 that is taken up by the thyroid, the size and geometry of the thyroid, the average energy of beta rays from the physical decay of I-131, and the biologic half-life of I-131 in the thyroid.

Appendix 6 of the NCI report gives a summary of the available data regarding weight of the thyroid (in children, adults, and fetuses), fractional uptake of I-131 to the thyroid, the biologic half-life of I-131, and the homogeneity of the distribution of dose within the thyroid. Information about population averages of the basic determinants of dose varies from good (for thyroid weight and uptake in adults) to poor (for uptake and biologic half-life in fetuses, for which models, but few data, are available for dosimetry analysis).

The NCI reports dose conversion factors as ranging from 1.3 mrad/nCi (3.5×10^{-7} Gy/Bq) for adults to 12-15 mrad/nCi (3.2×10^{-6} to 4.1×10^{-6} Gy/Bq) for infants. The variability of such fundamental factors as the size of the thyroid (perhaps a 20-fold variation around the median values in adults and children at any given age, Figure A6.1, NCI report) is large at the individual level. Greater

---

[2] The units of mrad/nCi as used in the NCI report are conventional units; the same quantity could also be expressed in international units of gray per bequerel (Gy/Bq) though the numerical value would be different (1 mrad/nCi = 2.7×10^{-7} Gy/Bq).

thyroid mass translates to lower dose (dose = energy absorbed per gram of tissue), so individual doses in response to the same intake of I-131 are potentially highly variable. (Differences in thyroid size between individuals could be offset by compensating factors in fractional uptake.) This likely interindividual variability is a source of additional skepticism concerning the factor-of-5 estimate of the variability of individual thyroid dose estimates cited in Chapter 8 of the NCI report.

The population of the United States between 1950 and 1960 was about 160 million, and the derived average thyroid dose from the I-131 produced by the collective tests (150 MCi or 5.6×10^{18} Bq) was estimated to be about 0.02 Gy (2 rad). It should be noted, however, that doses to specific persons can vary substantially about this mean, depending on age at exposure, diet, and, to a lesser extent, location at the time of the tests. To give perspective to the estimated average thyroid dose, this committee observes that standards for radiation exposure of the public evolved during the 1950s perhaps because of the presence of radioactive fallout from weapons testing in the United States and elsewhere. As developed in Appendix E of this report, the standards available during the Nevada Test Site operations were for protection of radiation workers, with limits for thyroid exposure that ranged from 15 to 30 rem per year. It became general practice in the 1950s to limit public exposure to one-tenth that for occupational workers, and on this basis exposures of 1.5 to 3 rem per year to members of the public would have been considered safe with no requirement for public intervention. In particular, children and other persons regularly consuming milk from backyard animals could have received doses that exceeded these limits. Athough the doses to most people over the age of 15 years were below the limits, it should be recognized that the limits applied to all individuals, not population averages.

VALIDATION AND UNCERTAINTY IN COLLECTIVE OR AVERAGE DOSE ESTIMATES

It is especially important, from the point of view of assessing the overall validity of the NCI report, to confirm the collective and average dose to the American people and the extent of the uncertainty in these estimates. This committee has, therefore, developed a "top down" approach to validate NCI's "bottom up" approach. It should be stressed that this committee's approach should not be seen as a substitute for the rigorous analysis applied by NCI, but it does provide the opportunity to confirm the general correctness of NCI's primary estimate.

Only iodine that undergoes radioactive decay in the thyroid glands of U.S. citizens contributes to the collective dose. If NCI's estimate of the amount of I-131 released in the total testing period (150 MCi) and the collective dose of 4×10^8 person-rad are correct, only 1 in 1 million I-131 decays contributes to the collective dose; the remaining 999,999 decay harmlessly on the ground, in the atmosphere, in cattle, and so forth. The committee therefore assumed that the pasture-cow-milk-man pathway was the dominant route of exposure and considered 6 "filtration" steps along the route through which those I-131 decays contributing to collective dose would have to pass. The steps are as follows:

- Step 1, the fraction of I-131 deposited on the mainland of the United States.
  - Step 2, the fraction retained on pasturage.
  - Step 3, the fraction consumed by cattle before it decayed.
  - Step 4, the fraction of total pasturage consumed by cattle.
  - Step 5, the fraction the cattle consumed and that entered the milk.
  - Step 6, the fraction in total produced milk that was consumed.

By then multiplying the amount of residual activity that remains after Step 6 by the total released activity, and converting that to dose, the collective dose of U.S. citizens is estimated. For example, of the amount released, 25 percent was estimated to be deposited in the 48 contiguous states and 75 percent either decayed in the atmosphere, was carried out of the contiguous states, was deposited on nonpasture land, or was otherwise filtered out of the chain. At Step 2, of the 25 percent deposited, the fraction intercepted by the pasture is estimated. This process continues through the remaining steps. Full details of the calculation are given in Appendix C.

At the completion of this "chain of filters" the estimated collective dose is 8 $\times 10^8$ person-rad, twice the value determined by NCI. Given the rough nature of this committee's calculations, this is considered by the committee to be good agreement and should provide confidence that the NCI estimate is not grossly under or over the actual value.

This committee is, however, concerned that NCI's estimate of the uncertainty in the collective dose is too low. The only discussion to date of this figure has been in the material presented to the committee by Dr. Charles Land of NCI. The material states that the GSD for the collective dose estimate is 1.4, so the per person thyroid dose could vary by a factor of 2 in each direction (that is, the average individual doses lie between 0.01 and 0.04 Gy). Based on estimates of the uncertainty in each step in the filtration chain above, the 5 percent and 95 percent confidence limits for the collective dose are 5×10^7 and 3×10^9. This would give the range for average individual doses as 0.0025-0.15 Gy (0.25-15 rad). In fact, to arrive at an uncertainty factor of 2 in either direction, using the simplified model above, would require an uncertainty of 30 percent at each step in the filtration chain. Given the overall effect of this filtration process, the committee is concerned that the NCI estimate of uncertainty is be underestimated.

The range of likely collective dose estimates implies that there is a probability (see Chapter 3 of this report) that substantial elevations in thyroid cancer incidence were produced by the Nevada tests for large groups of Americans. At the same time, there is a corresponding probability that the NCI report's calculation is too large and that the increase in thyroid cancer incidence is quite small.

Because of the sensitivity of the thyroid to radiation, epidemiologic methods (the study of observed thyroid cancer rates) could in fact provide an important indication of the size of the collective dose, and such analyses can and should be used to reduce the uncertainties in the collective dose estimate.

This committee, by independently validating the "order of magnitude" of the NCI's estimate of collective dose, is confident that there has been no gross over- or underestimation, but it is less confident that the uncertainty in the estimate has been realistically determined by NCI. It should, however, be noted that within these collective and average doses are concealed large differences in individual doses that depend on factors such as lifestyle and age at the time of testing. In this aspect lie important public-health implications of the Nevada atomic weapons tests.

Variation of Estimated Doses by Geographic Location and by Year of Birth

Considerable variation of possible doses is obvious from inspection of the data from the NCI report at a selection of locations across the country. Table 2.1 provides a summary of thyroid doses for 20 cities across the United States for four different consumption scenarios. In addition to the wide variation, it is obvious that some northeast locations have predicted doses similar in magnitude to those in the mountain states.

Considerable variation in dose is also predicted as a consequence of date of birth for a single location (Denver, CO is used as an example). Table 2.2 gives the estimated dose for individuals born in 5 year increments from 1937 to 1962. The date of birth resulting in greatest exposure is about 1 January 1952.

This table shows the relative difference in dose due to the age at time of testing. For example, an individual born in 1932 would have received 9 percent of the dose of someone born in 1952; someone born in 1947 would have received about one half that dose.

CONCLUSIONS

The NCI report presents a comprehensive rationale for assuming that significant thyroid doses were experienced by many in the U.S. population, particularly for the youngest birth cohorts, as a result of the fallout of I-131 from the nuclear weapons testing program. The overall report is thorough, but complicated, and complexity is the price paid for the amount of detail in the results presented, for example, in the estimates for representative age groups and for milk intake patterns in each county within the continental United States. Between-county and systematic uncertainties are important in determining the public-health significance of the estimates. Although it is recognized—both in the report and in the material made available on the NCI Web site—that uncertainties in county dose

TABLE 2.1 Thyroid Dose (cGy or Rad) for an Individual Born on January 1, 1952

Dietary source of iodine-131	Average diet with retail commercial milk				Average diet with milk from a "backyard cow"				Average diet and milk from a backyard goat				Average diet without milk			
	(95% Uncertainty Range)				(95% Uncertainty Range)				(95% Uncertainty Range)				(95% Uncertainty Range)			
Location (Co.)	2.5% tile	Geo. Mean	GSD	97.5% tile	2.5% tile	Geo. Mean	GSD	97.5% tile	2.5% tile	Geo. Mean	GSD	97.5% tile	2.5% tile	Geo. Mean	GSD	97.5% tile
(units)	(cGy)	(cGy)		(cGy)	(cGy)	(cGy)		(cGy)	(cGy)	(cGy)		(cGy)	(cGy)	(cGy)		(cGy)
Pacific Coast																
Los Angeles, CA	0.02	0.19	3.1	1.7	0.2	0.7	1.9	2	0.9	6.4	2.8	48	0.001	0.013	4.6	0.26
Alameda, CA	0.2	1.4	2.6	9.1	0.4	2.9	2.6	19	3.0	21	2.7	147	0.01	0.048	2.8	0.36
Coos, OR	0.4	2.0	2.2	9.4	1.0	4.0	2.0	16	4.0	22	2.4	122	0.01	0.081	4.1	1.3
Western States																
Navajo, AZ	0.3	2.8	2.9	23	4.9	21	2.1	169	20	120	2.5	723	0.09	0.48	2.4	2.7
Clark, NV	0.4	5.5	3.6	68	2.9	16	2.4	197	3.5	38	3.4	418	0.04	0.82	4.4	15
Washington, UT	2.5	23	3.1	211	9.2	51	2.4	468	13	210	4.1	3336	0.3	3.2	3.2	31
Mountain States																
Boise, ID	1.4	19	3.8	260	3.4	31	3.1	424	15	180	3.5	2097	0.05	1.5	5.8	47
Meagher, MT	2.5	43	4.3	750	3.8	55	3.9	959	20	330	4.2	5496	0.11	0.91	2.9	7.3
Denver, CO	4.2	12	1.7	34	11.5	29	1.6	82	34	120	1.9	422	0.04	0.52	3.5	6.1
Central States																
Scott, MN	2.7	15	2.4	83	4.5	23	2.3	128	18	120	2.6	781	0.11	0.61	2.4	3.4
Milwaukee, WI	2.2	8.4	2.0	33	2.8	13	2.2	51	20	79	2.0	307	0.08	0.31	2.0	1.2
St. Louis, MO	2.3	13	2.4	72	5.4	30	2.4	167	23	130	2.4	723	0.20	0.69	1.9	2.4
Franklin, KS	3.3	14	2.1	60	6.9	27	2.0	116	29	150	2.3	767	0.24	0.75	1.8	2.4

Southeast States

Orange, FL	**0.3**	1.8	2.4	**10**	**3.3**	14	2.1	**78**	**9.4**	61	2.6	**397**	**0.10**	0.25	1.6	**0.6**
Anderson, TN	**1.8**	6.5	1.9	**23**	**4.3**	15	1.9	**53**	**34**	120	1.9	**422**	**0.18**	0.46	1.6	**1.2**
Washington, DC	**1.2**	5.0	2.1	**21**	**1.2**	5.0	2.1	**21**	**13**	70	2.4	**389**	**0.02**	0.20	3.5	**2.3**

Northeast States

Albany, NY	**1.1**	9.7	3	**84**	**2.2**	23.0	3.3	**198**	**12**	100	2.9	**806**	**0.05**	0.43	2.9	**3.5**
Columbiana, OH	**0.18**	4.1	4.9	**92**	**0.23**	5.5	5.0	**124**	**1.1**	32	5.7	**970**	**0.02**	0.31	4.1	**4.9**
Hartford, CT	**1.5**	11	2.8	**83**	**3.4**	19	2.4	**143**	**14**	110	2.9	**887**	**0.07**	0.35	2.3	**1.8**
York, ME	**1.5**	7.7	2.3	**39**	**3.3**	13	2	**67**	**13**	75	2.4	**417**	**0.10**	0.31	1.8	**1.0**

NOTE: The original NCI Website (http:www2.nci.nih.gov/fallout/html) gives only the geometric mean and the geometric standard deviation. The 95% uncertainty range is calculated as 97.5%tile = $GM*(GSD^{1.96})$; 2.5%tile = $GM/(GSD^{1.96})$.

TABLE 2.2 Variations in Dose as a Consequence of Date of Birth

Birthdate	Reference dose (mGy)^a	Ratio of dose in birth year to dose for those born in 1952
Jan. 1, 1962	0.0001	0.000001
Jan. 1, 1957	85	0.71
Jan. 1, 1952	120	1.00
Jan. 1, 1947	64	0.53
Jan. 1, 1942	44	0.37
Jan. 1, 1937	25	0.21
Jan. 1, 1932	11	0.09

^{a}Divide by 10 to convert to rad.

estimates are important, no clear statements in the report distinguish between systematic and between-county uncertainties. It is likely that within-county uncertainties produce a significant portion of overall uncertainty in thyroid dose, which means that the level of detail presented (county estimates, rather than estimates over much broader regions) is probably inappropriate.

3

Health Risks of I-131 Exposure

The major health risks associated with exposure to iodine-131 (I-131) involve the thyroid gland, which concentrates this radionuclide. Assessment of the magnitude of the public-health problem posed by exposure to I-131 estimated by the National Cancer Institute (NCI 1997a) entails understanding

- The biology of the thyroid gland.
- The relationship of exposure to ionizing radiation and the occurrence of thyroid cancer.
- The effect of radiation on the frequency of nonmalignant thyroid disease.
- Projections of the risk of thyroid cancer through the lifetime of exposed individuals.
- The estimates of the proportion of cases of I-131 related thyroid cancer that have already occurred.

THYROID GLAND BIOLOGY

The thyroid gland (see Figure 3.1) is a butterfly-shaped, ductless gland astride the trachea on the anterior side of the throat.

The gland usually begins as an endodermal thickening and a pouch in the floor of the pharynx, visible about 3 weeks after conception. Thyroid follicular cells develop in the embryo, and by the 10th week of gestation, iodine is accumulated and colloid is present within the follicles. Thyroxine then becomes detectable and the gland is functional (O'Rahilly and Muller 1992). The thyroid gland is the source of several hormones in which iodine is an important constituent. The thyroid is the only organ in the body that greatly concentrates and retains iodine.

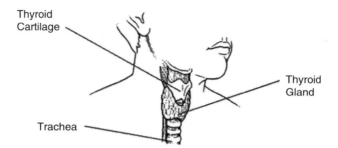

FIGURE 3.1 Anatomical drawing of thyroid location (courtesy of American Cancer Society).

Normal Thyroid Physiology

Concern about the carcinogenic effects of exposure to radioiodine on the thyroid gland is motivated by three major factors. First, evidence has accumulated that the thyroid gland is uniquely sensitive to the effects of radiation. There is some evidence that measurable increases in thyroid cancer can occur with external doses of radiation as low as 0.1 sievert (Sv) (10 rem). A finite risk at low doses of that magnitude is consistent with risks for other solid cancers reported for the Japanese atomic-bomb survivors (Pierce and others 1996). Second, the cow-milk-man pathway described in the NCI (1997a) report and discussed in Chapter 2 of this report provides a mechanism by which radioiodines in the environment can be greatly concentrated in the human food chain. Finally, because most of the radiation dose is from ingested or inhaled radioiodine, the radiation dose to the thyroid is 500-1,000 times greater than is the largest radiation dose to other organs in the body.

For several reasons, persons exposed to I-131 as children are uniquely at risk for carcinogenic effects. First, children drink more milk relative to their body size than do adults. Second, the same amount or a higher fraction of internalized iodine is concentrated in the smaller thyroid glands of children; therefore the radiation dose to the thyroid in children is higher than it is in adults. Finally, studies of children whose thyroid glands were exposed to external radiation suggest a strong inverse relationship between age at exposure and the carcinogenic effects of radiation on the thyroid. Over the age of 15, little increase in thyroid cancers has been observed. Below the age of 15, thyroid cancer increased by a factor of approximately 2 for every 5 years' decrease in age. Not only is the frequency of malignant nodules increased by thyroid irradiation, but benign nodules also occur with greater than usual frequency after irradiation (Wong and others 1996).

Stable iodine and its radioactive isotopes are water-soluble and readily absorbed, either from the gastrointestinal tract after ingestion or through the lungs after inhalation. The first step in the synthesis and storage of thyroid hormone involves a mechanism for concentrating iodine from extracellular fluid, variably called the iodine pump, the transport mechanism, the iodide-concentrating mechanism, or the iodine trap. Transport of I^- across the thyroid membrane is an energy-dependent process linked to the transport of sodium; this fact led to the concept of an Na^+-I^- cotransport (symport) system, with an ion gradient generated by Na^+-K^+ ATPase as the driving force. By this mechanism, the thyroid attains remarkably high concentrations of iodide; concentrations of 30 to 40 times that in blood are usual though values in excess of 400 fold over the level in the bloodstream have been recorded (Taurog 1996). Other tissues in humans contain sodium iodide symporters: the gastric mucosa, salivary glands, mammary glands, choroid plexus, ovaries, placenta, and skin (Smanik and others 1996). Breast tissue, which contains iodine symporters, can therefore pump iodine into breast milk.

Once iodine is concentrated in the thyroid follicular cell, it is incorporated into tyrosine molecules that form part of a larger protein, thyroglobulin. Thyroglobulin is the storage form of thyroid hormone that is kept, often for long periods, within the thyroid gland. Once iodine has been incorporated into proteins by the thyroid, the biologic half-life of iodine within the thyroid is typically 80-120 days; non-protein-bound iodine has a biologic half-life of several hours in the body. The long half-life of thyroid iodine results in nearly all of the energy from the I-131 being deposited in the thyroid. The liver inactivates thyroid hormone, breaking it into smaller, biologically inert components that are eventually excreted by the kidney. Thyroid hormone is essential to life. It regulates many metabolic processes, including the rate of cellular oxygen consumption, and it affects the performance of many body systems, including the heart and nervous systems.

Breaking down thyroglobulin within the thyroid produces two main forms of thyroid hormone, tetraiodothyronine and triiodothyronine, which are then secreted into the blood. Tetraiodothyronine (thyroxine) is secreted in much greater quantities than is triiodothyronine; it has 4 iodine molecules and a half-life of about 7 days in the circulation. Triiodothyronine, the most potent thyroid hormone, has 3 iodine molecules and a half-life of about 12 hours in serum. Most of the triiodothyronine in the blood comes from conversion of tetraiodothyronine to triiodothyronine by the body.

The unique ability of the thyroid gland to concentrate iodine has enabled the effective use of radioiodines in the diagnosis and treatment of thyroid disorders, including an overactive thyroid (Graves disease or toxic multinodular goiter), and differentiated (papillary and follicular) thyroid cancers (Mazzaferri and Jhiang 1994). Given for medical purposes in doses that range from 5 to 200 millicuries (mCi), I-131 efficiently destroys overactive and malignant thyroid tissues.

For many years, I-131 was used in very small amounts (50-100 mCi) for diagnostic studies. Typically, these were 24-hour thyroidal radioactive iodine uptake, which is a measurement of the amount of iodine taken up by the thyroid from the blood and thyroid imaging studies that give some information about the configuration of the thyroid. (This is in contrast to larger doses of I-131, in the range of 10 to 200 mCi, that are given to ablate malignant thyroid tissue or to treat overactive thyroid glands. Large doses of I-131 ordinarily destroy the thyroid gland and thus do not induce thyroid cancer.) In addition to diagnostic and research exposure, children have also experienced therapeutic exposure to I-131 as described below.

Thyroid Cancer and Thyroid Nodules

Thyroid cancer is usually clinically manifested as a nodule on the gland. Most thyroid nodules are benign. Palpable thyroid nodules, both benign and malignant, increase in frequency with age and are more common among women than they are in men. Although studies vary, perhaps 5 percent of women over the age of 50 and about 1 percent of men over 50 have thyroid nodules that can be felt during physical examination.

The prevalence of thyroid nodules detected by ultrasonography is as much as 10-fold greater than the prevalence of palpable thyroid nodules, depending on the population (Tan and Gharib 1997; Ezzat and others 1994). Most thyroid nodules detected by ultrasound are small (<1 cm in diameter) and not palpable, whether or not the population being studied has received thyroid radiation (Schneider and others 1997; Tan and Gharib 1997; Ezzat and others 1994).

Larger thyroid nodules (1.5 cm or larger) are more likely to be associated with clinically significant thyroid cancer (Mazzaferri and Jhiang 1994). For several reasons, however, even these large nodules are not always palpable. First, to detect a nodule by palpation, its consistency must be recognizably different from the consistency of the normal thyroid gland. Second, some nodules are in areas that are difficult to palpate, such as on the back surface of the gland or behind the sternum. Third, the thickness of the neck of some patients makes examination of the thyroid difficult. Finally, the examiner's skill and the completeness of the examination will, in part, determine the palpability of the nodule. In one study of 54 individuals who had been exposed to therapeutic head and neck irradiation during childhood for benign conditions, ultrasound detected 157 nodules in 87 percent (47) of the subjects; 52 percent (28) had 40 nodules in all that were 1.0 cm or larger. Of the 11 nodules that were 1.5 cm or larger, palpation detected only 5, or 45 percent (Schneider and others 1997). Other studies of populations not exposed to I-131 have reported better results (Chapter 4).

Most thyroid nodules biopsied by fine-needle aspiration (FNA) are benign, even in patients who have a history of head and neck irradiation (Mazzaferri 1993a). The high prevalence of benign thyroid nodules in the general population and among persons with a history of head and neck irradiation increases the risk of false-positive test results. This is discussed in detail in Chapter 4 of this report.

Incidence of Clinically Manifest and Occult Disease

The two principal malignancies of the thyroid follicle cell are papillary and follicular thyroid cancer. Malignant tumors resulting from exposure to ionizing radiation are almost exclusively papillary cancers (Nikiforov and Gnepp 1994). Those tumors also account for more than 80 percent of the thyroid cancers occurring spontaneously among persons with no known history of thyroid radiation (Mazzaferri 1991). According to American Cancer Society estimates, 17,200 new cases of thyroid cancer will be diagnosed in 1998 in the United States, ranking thyroid cancer 14th in incidence among 35 categories (Figure 3.2). Its incidence varies with gender and age and is highest in women between the ages of 30 and 70 years; the peak incidence reaches 13.2 per 100,000 per year between the ages of 50 and 54 (see Table 3.1).

The incidence of thyroid cancer is lower in men. In men, thyroid cancer peaks between the ages of 60 and 70, when its annual incidence is 8.6 per 100,000 (NIH 1997). In the latest Surveillance, Epidemiology, and End Result report (SEER 1998), the average lifetime risk over a 95-year lifespan of being diagnosed with some form of thyroid cancer was 0.66 percent (6.6 per 1,000) for

TABLE 3.1 Thyroid Cancer (Invasive) Incidence Rates per 100,000 Persons, 1990+1994, by Age at Diagnosis

Age at Diagnosis	Total	Males	Females
All ages	4.9	2.8	6.9
0-4	0.0	0.0	0.0
5-9	0.1	0.1	0.1
10-14	0.4	0.3	0.6
15-19	1.4	0.3	2.6
20-24	3.9	1.0	6.8
25-29	5.4	2.3	8.6
30-34	6.8	2.4	11.1
35-39	7.8	3.7	11.8
40-44	7.9	3.7	12.0
45-49	8.5	4.8	12.2
50-54	9.5	5.7	13.2
55-59	9.3	6.1	12.4
60-64	8.6	7.0	10.1
65-69	9.9	7.5	11.8
70-74	9.7	8.6	10.5
75-79	9.5	8.5	10.2
80-84	7.7	6.1	8.6
85+	7.9	7.2	8.2

SOURCE: (NCI 1997b).

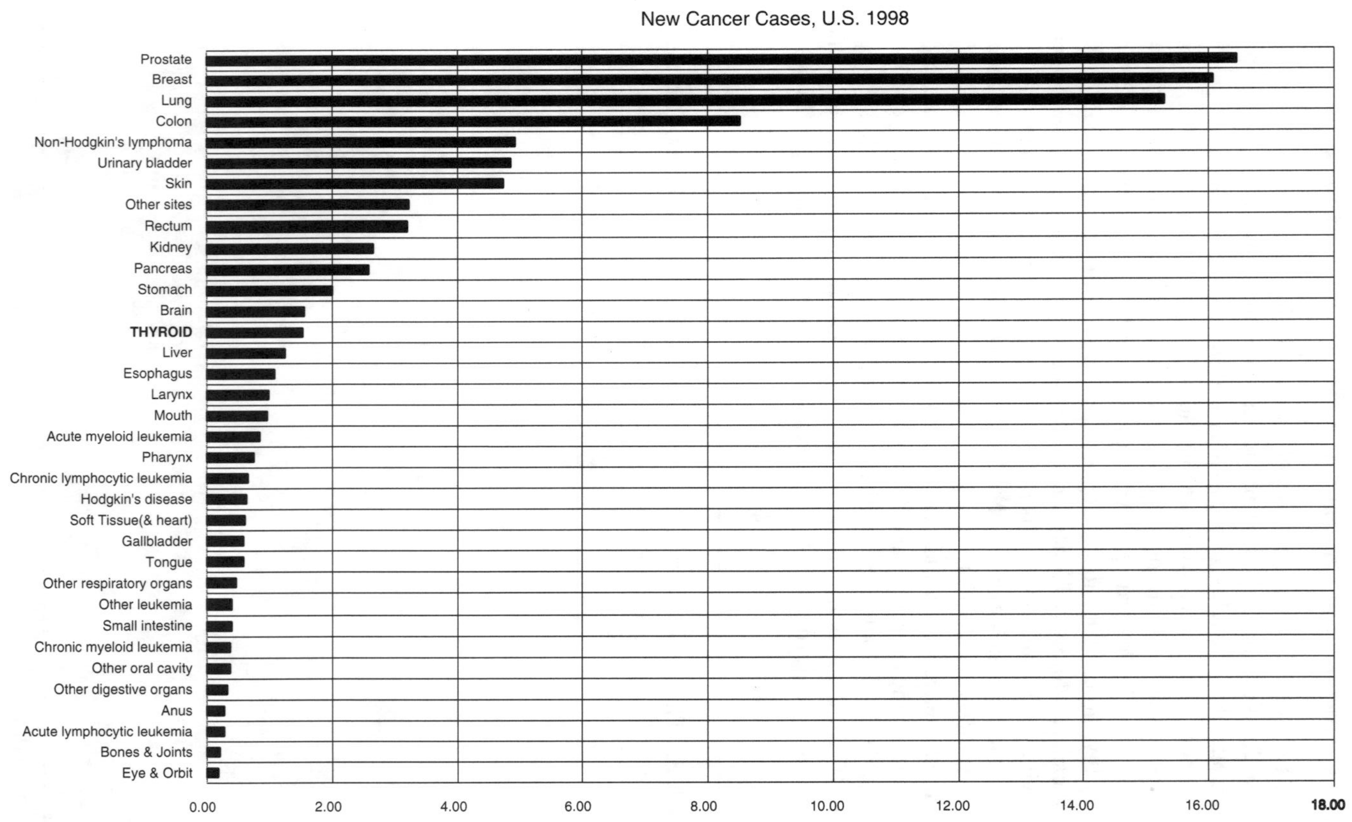

FIGURE 3.2 Percent of all cancer cases (data from ACS 1998).

women and 0.27 percent (2.7 per 1,000) for men. By way of comparison, the lifetime risks for women of developing invasive in situ breast cancer or invasive in situ colon cancer are 14.2 percent and 6 percent, respectively. For men, the lifetime risks of developing prostate or lung cancer are 18.8 percent and 8.4 percent, respectively. The lifetime risk of dying from some form of thyroid cancer was 0.07 percent for women and 0.04 percent for men. These figures compare to 3.46 percent and 2.53 percent lifetime risks of dying, respectively, from breast cancer and colon/rectal cancer for women and 3.64 percent and 2.57 percent for prostate and colon/rectal cancer for men.

The risk of dying of thyroid cancer in countries with efficient medical care systems is low. The long-term mortality rates for papillary thyroid carcinoma are less than 10 percent at 30 years after diagnosis (Mazzaferri and Jhiang 1994). The American Cancer Society estimates that 1,200 people will die from thyroid cancer in 1998, accounting for about 0.2 percent of all cancer deaths. Unlike its incidence, which has been rising, the mortality rates for thyroid cancer have been falling. Between 1973 and 1994, the mortality rates for thyroid cancer dropped more than 23 percent, both for people younger than 65 years and for people older than 65 at the time of diagnosis (NIH 1997). See Figure 3.3.

Between 1973 and 1992, the incidence of thyroid cancer rose almost 28 percent ($p < 0.05$)—a change that has been observed in persons both under and over the age of 65 at the time of diagnosis. In the SEER reports, 14 of 23 cancer sites showed increasing incidence during this period; only 4 of the 14, including thyroid cancer, showed decreasing mortality. The contrast between the incidence and mortality trends has been attributed to more sophisticated detection technologies (ultrasound for nodules and FNA biopsy for cancer) and more complete diagnostic reporting (Wang and Crapo 1997).

A large number of thyroid cancers are small, occult tumors that are usually not detected during a person's lifetime and that rarely progress to cause problems. Clinically silent tumors are generally papillary microcancers smaller than 1.0 cm in diameter. They may be found unexpectedly during surgery for benign thyroid disease, at autopsy, or by FNA biopsy of a nodule discovered by ultrasonography. Their prevalence varies according to the geographic location and possibly ethnicity, the type of tumor, and the intensity of the pathologic examination (Moosa and Mazzaferri 1997). In autopsy studies of persons who died without known thyroid disease, the prevalence of occult thyroid cancer ranges from 5 to 13 percent among studies in the continental United States and 6 to 36 percent among studies in Europe (Moosa and Mazzaferri 1997; Thorvaldsson and others 1992; Harach and others 1985). Occult cancer is found in all age groups but is more frequent after the age of 40; there is no gender difference in frequency. Thus, there is good reason to suspect many healthy people harbor tiny thyroid cancers that will never harm them.

The problem of microcancers is not unique to the thyroid gland. Similar tumors are found even more commonly in the breast and prostate. The introduction

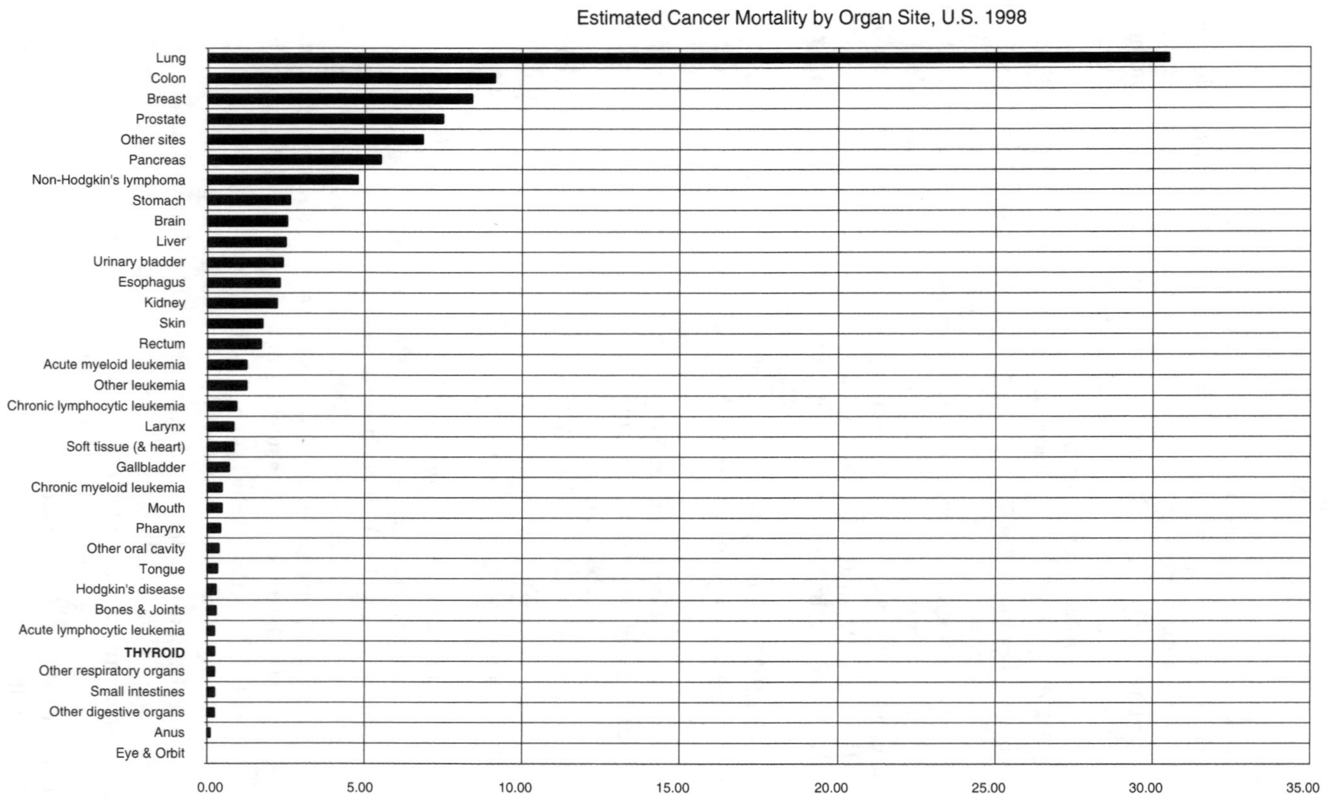

Estimated Cancer Mortality by Organ Site, U.S. 1998

FIGURE 3.3 Percent of all cancer mortality (data from ACS 1998).

of sophisticated diagnostic tests has resulted in the discovery of many microcancers that are unlikely to harm the patient. A major challenge for medical research is to differentiate clinically significant microcancers from those that will never harm the patient. Failure to make this differentiation will result in some patients undergoing treatment for harmless diseases and others imprudently having their diseases ignored. Research cited in Chapter 4 suggests that people want to factor information about these usually nonprogressing cancers into their decisions about cancer screening and treatment.

Thyroid Cancer in Persons (All Ages) Not Exposed to Radiation

Most clinically apparent papillary thyroid cancers are first manifested as one or several palpable thyroid nodules, discovered in about half the cases by the patient (Mazzaferri 1993a). They are otherwise usually asymptomatic, although a small proportion of highly invasive tumors are very symptomatic. At the time of diagnosis, the primary tumor is typically 2.0-2.5 cm, but can range from a few millimeters to more than 5 cm in diameter.

With routine study of permanent histologic sections, about 20 percent of papillary cancers are multiple tumors thought to represent intrathyroidal metastases, but with meticulous study more small tumors (up to 80 percent in some studies) are usually apparent within the gland (Mazzaferri 1991). Some 5-10 percent of the primary tumors that occur without known exposure to radiation invade the thyroid capsule, growing directly into surrounding tissues, thus increasing both the morbidity and the mortality of papillary cancer (Mazzaferri and Jhiang 1994; Emerick and others 1993). The most commonly invaded structures are the neck muscles and vessels, recurrent laryngeal nerves, larynx, pharynx, and esophagus—but tumors can extend into the spinal cord and brachial plexus. At the time papillary cancer is diagnosed, about 40 percent of adult patients have metastases to regional lymph nodes and about 5 percent have distant metastases, usually to the lung (Mazzaferri 1991).

Mortality rates for adults with papillary thyroid cancer are generally less than 10 percent over several decades after initial therapy (Mazzaferri 1993b). Cancer-specific mortality rates in adults with papillary cancer are about 5 percent at 10 years and slightly less than 10 percent at 20-30 years after treatment; the 5-year survival rate is only about 50 percent for patients with distant metastases (Dinneen and others 1995; Mazzaferri and Jhiang 1994; Mazzaferri 1991; Hay 1990). As is characteristic of many cancers and other diseases, cancer-specific mortality rates are progressively higher for patients over age 40 (Figure 3.4) and among persons with more advanced tumor stages at the time of diagnosis.

Thyroid Cancer in Children Not Exposed to Radiation

Thyroid cancer that occurs spontaneously has somewhat different features in

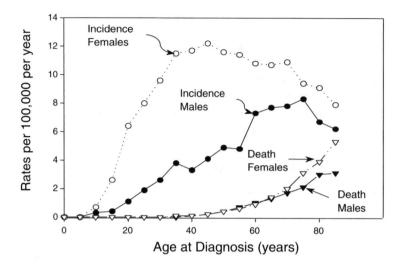

FIGURE 3.4 Incidence and cancer-specific mortality rates for thyroid carcinoma. Drawn from the data published by Kosary CL et al. 1995. SEER Cancer Statistic Review, 1973-1992: Tables and Graphs. National Cancer Institute. NIH Pub. No. 96-2789, Bethesda, MD.

young children than it has in adults. In children, it is almost always papillary and usually is at a more advanced stage at the time of diagnosis. Papillary cancer in children more frequently invades beyond the thyroid capsule, and it metastasizes to regional lymph nodes in almost all cases (Hung 1994; Robbins 1994; De Keyser and Van 1985). For example, in a study of 98 children with differentiated thyroid cancer (Travagli and others 1995), lymph node involvement was seen in 88 percent of children at the time of diagnosis, and invasion of the thyroid capsule had occurred in 59 percent.

Distant metastases also are more frequent in children than they are in adults with differentiated thyroid cancer. In some series, up to 20 percent of children have distant metastases at the time of diagnosis (about 4 times the rate that occurs in adults) and another 10-20 percent of children develop them during the course of the disease (Harness and others 1992; Schlumberger and others 1987; Goepfert and others 1984). In fact, distant metastases are most frequently observed in the youngest patients, especially those who are younger than 7 at initial treatment. There is a high recurrence rate in children after initial surgical removal of tumors.

Despite the aggressiveness of thyroid cancer in children, the long-term mortality rate is only about 2.5 percent, so survival is thus much better for children than it is for adults (Figure 3.2) (Robbins 1994). Because death from recurrent disease can occur many years later, however, the prognosis evolves over decades. Moreover, the statistics are somewhat misleading. In the study of Travagli and others (1995), although relatively few deaths occurred in children, the standardized mortality rate (SMR) was 6.4; however, 95 percent confidence intervals (CIs) were not reported, and a few deaths might result in an unrealistically high SMR.

Ret Proto-Oncogene and Papillary Thyroid Cancer

Our understanding of the molecular genetics of thyroid cancer has grown substantially in recent years (Fagin 1994b; 1994a; Farid and others 1994). Of particular interest in patients with papillary thyroid carcinoma, and especially in children who have been irradiated, are the genes on chromosomes 10 and 17 involved in paracentric inversions or translocations that result in the activation of the tyrosine kinase domain of the *ret* proto-oncogene. This is the most common event in papillary thyroid cancers occurring naturally (PTC1) and among those in children after the Chernobyl accident (PTC3).

Normally, *ret* is not expressed in thyroid follicular cells and its promoter is thus inactive. In papillary thyroid cancer, but not in other thyroid neoplasms, the tyrosine kinase domain of *ret* is turned on and activated by a paracentric inversion on chromosome 10 involving *ret* and another gene, H4, producing PTC1 (papillary thyroid cancer 1) (Grieco and others 1990). Two other genes are similarly rearranged with *ret*: RI, which codes for a subunit of the receptor-associated Gs protein that forms PTC2 (Santoro and others 1994), and ELE1, to form PTC3 or PTC4 (Fugazzola and others 1996; Klugbauer and others 1996; Jhiang and others 1994).

Ret proto-oncogenes have been detected in 11-59 percent of naturally occurring human papillary thyroid cancers, depending on the means of detection and the population studied (Williams and Tronko 1996). The most common rearrangement among patients with sporadic tumors is PTC1 (Jhiang and Mazzaferri 1994), while PTC3 is the most common in children from the area around Chernobyl who developed thyroid cancer.

RADIATION AND THYROID CANCER

Thyroid Cancer from External Radiation Exposure

Studies evaluating the risk of thyroid cancer from radiation exposure have credence insofar as they use reasonably accurate dosimetry (calculation of radiation doses to the thyroid), have substantial numbers of persons in the dose range of interest (for this population, low to moderate doses), have a reasonably long follow-up period, and have a high follow-up rate. The statistical power and precision of such studies, which are important in weighing the study results, are positively related to the number of thyroid cancers observed and to the mean dose.

Summary results of the seven principal cohort studies of thyroid cancer incidence among those irradiated externally before the age of 20 years are given in Table 3.2. The study of Japanese atomic-bomb survivors has an appreciable number of subjects in the low-to-moderate dose range, as do several medical irradiation studies. Some studies include people who received doses over a period of time rather than during a single episode.

This review concentrates on data concerning irradiation before age 20 and its long-term consequences. In the Japanese atomic bomb study (Thompson and others 1994), a strong effect of age at exposure on thyroid cancer incidence was seen, such that the excess relative risk (ERR) per Sv was 9.5, 3.0, 0.3, and −0.2 at ages 0-9, 10-19, 20-39, and 40+ y, respectively. (See Glossary for an explanation of ERR and other technical numbers in this chapter.) This provides compelling evidence that thyroid cancer risk is inversely related to age at irradiation and that there is little, if any, cancer risk from irradiation after age 20. The studies of radiotherapy for *tinea capitis* (ringworm of the scalp) in Israel (Ron and others 1989) and for enlarged tonsils in Chicago (Schneider and others 1993) also reported inverse relationship between age at exposure and cancer risk, although these studies had more restricted age ranges.

In the study of cancer incidence among Japanese following atomic-bomb

TABLE 3.2 Thyroid Cancer ERR and EAR for Cohort Studies with Acute External Irradiation before Age 20

Study (Reference)	Mean Dose (Gy)	Observed/ Expected Cancers	ERR per Gy (95% CI)[a]	EAR per 10^4 person-year Gy[a]
A-bomb (<15 y at exposure) (Ron and others 1995; Thompson and others 1994)	0.26	40/ 19.2	4.7 (1.7,11)	2.7
Enlarged thymus (Shore and others 1993)	1.36	37/ 2.8	9.1 (3.6,29)	2.6
Tinea capitis (Ron and others 1989)	0.09	44/ 11.2	32.5 (14,57)[b]	7.6
Enlarged tonsils (Schneider and others 1993)	0.59	309/ 125	2.5 (0.6,26)	3.0
Lymphoid hyperplasia (Pottern and others 1990)	0.24	13/~2.2[c]	~20 (9.5,37)	15.1
Skin hemangioma (Lundell and others 1994)	0.26	17/ 7.5	4.9 (1.3,10)	0.9
Skin hemangioma (Lindberg and others 1995)	0.12	15/ 8.0	7.5 (0.4,18)	1.6

[a] Both ERR and EAR estimates were based on dose-response analyses.

[b] When an indicator variable for irradiated vs. control group was included, the ERR slope dropped to 6.6 per Gy.

[c] Value estimated for this tabulation from the data available.

exposure (Thompson and others 1994), 132 thyroid cancers were observed among those with thyroid doses less than 10 mSv. There was a linear dose-response relationship ($p < 0.001$) with no significant nonlinearity ($p = 0.17$). As noted above, the excess risk caused by radiation was confined mainly to those under age 20 at the time of the bomb; the ERR was 0.10 (95 percent CI, −0.23 to 0.75) among those irradiated at age 20 or above. Among those under age 20 at irradiation, there were 59 thyroid cancers in the group receiving 10 mSv and 25 among those with <10 mSv. The excess risk was statistically significant for ages 0-9 and 10-19 at irradiation. The background incidence was about 3 times as high for females as for males and it was 2.5 times as high among those who received biennial examinations in the Adult Health Study (AHS) than it was among those who did not; but the radiation dose-response slopes were similar by gender ($p > 0.5$) and AHS status ($p > 0.4$).

A study in Israel of 10,834 children x-irradiated for *tinea capitis* found 43 thyroid cancers (RR, 4.0; 95 percent CI, 2.3-7.9) (Ron and others 1989). A dosimetric study for this group showed the average dose was about 0.09 Gy (9 rad) (Werner and others 1968), which has been supported by two other studies (Harley and others 1976; Lee and Youmans 1970). The 1968 study contributes strong evidence for an effect at a relatively low dose. A much smaller study of patients irradiated for scalp ringworm found no substantial excess of thyroid cancer (2 observed, 1.3 expected), but the two studies are marginally compatible statistically ($p = 0.07$ for the difference in risks after adjusting for gender and dose differences) (Shore 1992).

A Chicago study of 2,634 patients who received x-rays for enlarged tonsils showed a statistically significant excess of thyroid cancer after a mean dose of 0.6 Gy (60 rad) (based on 309 cancers) (Ron and others 1995; Schneider and others 1993). Follow-up in this study averaged 33 years. It is the only cohort study of radiogenic thyroid cancer that has included repeated thyroid screening over a period of years. Study limitations include the lack of an unexposed control group with a comparable intensity of screening and uncertainties in the thyroid doses.

A smaller study of x-irradiation for lymphoid hyperplasia (Pottern and others 1990) had a 29-year follow-up of 1,590 irradiated patients and a thyroid examination program. It showed an excess of thyroid cancer (13 cases) after an average thyroid dose of 0.24 Gy (24 rad).

A cohort of 2,657 infants x-irradiated for enlarged thymus gland has been followed an average of 37 years (Shore and others 1993). The mean dose was about 1.4 Gy (140 rad), but 56 percent had doses <0.5 Gy (50 rad). There was a strong dose-response association over the dose range. A statistically significant dose-response association was found even when the dose range was limited to <0.3 Gy (30 rad), but the limited statistical power precluded seeing a dose-response association at <0.2 Gy (20 rad). Thyroid cancer risk was elevated out to at least 45 years after exposure.

In Stockholm, Sweden, a cohort of 14,351 infants who were treated (mostly) with radium-226 for skin hemangiomas has been followed up for an average of 39 years by the Swedish tumor registry (Lundell and others 1994). The mean age at treatment was 6 months. The dose ranged from <0.01 Gy (1 rad) to 4.3 Gy (430 rad) with a mean of 0.26 Gy (26 rad). There was an elevated risk of thyroid cancer (standardized incidence rate [SIR], 2.28; 95 percent CI, 1.3-3.7) based on 17 cancers. The thyroid cancer excess persisted at least 40 years after exposure.

A study was conducted in Gothenburg, Sweden, of 11,807 infants treated with Ra-226 for hemangiomas of the skin and followed up for an average of 31 years by the Swedish tumor registry (Lindberg and others 1995). The median age at treatment was 5 months. The mean estimated thyroid dose was 0.12 Gy (12 rad). An excess of thyroid cancer (SIR, 1.88; 95 percent CI, 1.05-3.1) was found based on 15 thyroid cancers. One limitation of that study and the study of Lundell and others (1994) is that thyroid cancers were not ascertained until 1958, when the Swedish tumor registry began, even though some of the patients were treated as early as the 1920s. Hence, some thyroid cancers were probably never included.

A case-control study of thyroid cancer nested within a cohort study of second malignant neoplasms among childhood cancer survivors has been reported by Tucker and others (1991). There was an excess of thyroid cancer, but many of the thyroid doses were greater than 10 Gy (1000 rad). The dose-response curve plateaued, apparently because of cell killing related to the high exposures; thus, the risk estimate is of questionable applicability to low-exposure studies.

Ron and others (1995) conducted a pooled analysis (based on the raw data) of 5 of the major cohort studies of thyroid cancer among those given external radiation at less than 15 years of age. The combined data included 436 thyroid cancers. The pooled ERR was 7.7 per Gy (95 percent CI, 2.1-28.7). The wide CI occurred because of heterogeneity among the studies in risk estimates, which necessitated a random-effects model; the corresponding fixed-effects model yielded 95 percent CI of 4.9-12.0. The pooled excess absolute risk (EAR) was 4.4 per 10^4 person-year Gy (95 percent CI, 1.9-10.1). The ERR was marginally ($p = 0.07$) higher among females than among males. It peaked about 15 years after exposure but was still elevated 40 or more years after exposure. The ERR after fractionated exposure was about 30 percent less than after a single exposure, but the difference was not statistically significant. Although the test for curvilinearity was not statistically significant, the data suggested that a linear fit somewhat underestimated the risk at lower doses and overestimated it at higher doses.

Several case-control studies have been performed to determine the effects of medical diagnostic irradiation on thyroid cancer rates (Hallquist and others 1994; Ron and others 1987; McTiernan and others 1984). Of these, only one (Inskip and others 1995) used objective information rather than patient reports of diagnostic irradiation, with their potential for recall bias. Inskip's group found no association between thyroid cancer and the number of x-ray examinations of the head, neck, and upper spine (trend, $p = 0.54$) or the number of examinations of the chest, shoulders, and upper gastrointestinal tract ($p = 0.50$), nor was there an association for diagnostic x-ray examinations before 1960, when doses were probably much higher.

Thyroid Cancer from Exposure to Radioactive Iodine

The association between exposure of the thyroid gland to external ionizing radiation and the development of thyroid cancer is well documented for young children but not for older children or for adults (Ron and others 1995; Shore and others 1993). Whether internal radiation to the thyroid from radioiodine causes thyroid cancer in humans was, until recently, less certain, although it has been long recognized to induce thyroid cancer in animals (NCRP 1985; Lindsay and Chaikoff 1964). There is now strong evidence from Chernobyl that children exposed to radioiodine develop thyroid cancer at higher than usual rates.

Studies Other than Chernobyl

Table 3.3 summarizes information from studies of diagnostic, therapeutic, and fallout exposure to I-131. Hall and others (1995) have followed 34,104 patients who were administered I-131 for diagnostic purposes, but only a small

TABLE 3.3 Thyroid Cancer Excess Relative Risk (ERR) and Excess Absolute Risk (EAR) following Exposure to Iodine-131 before Age 20

Study (Reference)	Mean Dose (Gy)	Observed/Expected Cancers	ERR per Gy (90% CI)	EAR per 10^4 Person-Year Gy
Swedish Diagnostic [131]I (Hall and others 1995)	1.5	2/1.4[a]	0.3 (<0-2.7)	0.2
FDA Diag. [131]I (Hamilton and others 1989)	~0.8	4/1.4[b]	2.3 (<0-23)	0.4
Utah [131]I Fallout (Kerber and others 1993)	0.098	8/5.4	7.9 (<0-16)	2.1
Marshall Islands[c] (Robbins and Adams 1989)	12.4[d]	6/1.2	0.3 (0.1-0.7)	1.1
Juvenile Hyperthyroidism[e]	~88	2/ 0.1	0.3 (0.0-0.9)	0.1

[a]Based on patients whose ^{131}I examination was not for suspicion of thyroid tumor. One additional case was detected among those who had been examined for possible thyroid tumor (total observed/ expected = 3/1.8).
[b]The result given is based on their relatively small control group. The expected value based on population rates was 3.7, and the ERR was 0.1 (<0-2.0) with EAR = 0.05.
[c]Highly exposed group of 235 persons only.
[d]Over 80% of this dose was from short-lived radioiodines and external radiation rather than [131]I.
[e]Composite of 9 studies.

percentage of those patients were under age 20 at the time of exposure. About 10,800 were being examined for suspicion of thyroid tumor, and they showed a subsequent excess of thyroid cancer. Among those not being examined for suspicion of thyroid tumor, the average thyroid dose was about 0.7 Gy (70 rad), and no excess thyroid cancer was subsequently found (standardized incidence rate [SIR], 0.75; 95 percent CI, 0.5-1.1). Of particular interest was the subset of 1,764 patients exposed before 20 years of age but not in evaluation for suspected tumors, for whom the mean dose was about 1.5 Gy (150 rad). Among this group there were 2 thyroid cancers (SIR, 1.38; 95 percent CI, 0.2-5.0). It should be noted that fewer than 400 of these subjects were exposed before age 10, whereas the external radiation studies included many subjects exposed in the first decade of life.

A study conducted by the U.S. Food and Drug Administration (Hamilton and others 1989) reported a small excess of thyroid cancer among 3,503 juveniles given diagnostic I-131. The thyroid doses ranged from <0.1 (10 rad) to >10 Gy (1000 rad), with a median dose of about 0.35 Gy (35 rad) and a mean dose of about 0.8 Gy (80 rad). When compared with the unirradiated control group of 2,594 patients under other diagnostic modalities, there appeared to be a small excess of thyroid cancer (observed/expected, 4/1.4, nonsignificant), but when compared with general population rates there was no excess (4/3.7). One uncertainty in the study is the question of whether some of the diagnostic I-131 procedures were performed because of a suspicion of thyroid tumor.

Results for the total number of juvenile patients from several studies of I-131 therapy for hyperthyroidism are shown in the last line of Table 3.3. The excess of subsequent thyroid cancer was not statistically significant in any of these studies, which suffer from having rather loosely defined cohorts and follow-ups of variable quality.

In addition to studies of medical exposure to I-131, several studies have looked at those exposed as a result of exposure to radioactive fallout from nuclear weapons tests. One early concern about I-131 exposure resulted from the BRAVO hydrogen bomb test in the Marshall Islands in March 1954. The inhabitants of the northernmost atolls were exposed when the wind direction changed unexpectedly. A total of 253 persons were on 3 heavily exposed atolls (Rongelap, Ailinginae, and Utirik). Recent reports (Cronkite and others 1997; Howard and others 1997) indicate that 10 clinically significant thyroid carcinomas (plus 7 occult carcinomas) have been detected in this population, in comparison with 2 thyroid cancers found in a group of 227 unexposed persons. However, this might not be a valid comparison because in the unexposed group "some have not been seen for many years; others were added as recently as 1976" (Howard and others 1997). No thyroid surgery has been performed on this group since 1985 (the exposed group included surgery through 1990). Results in the unexposed group were not broken down by age so as to permit a comparison of those exposed before adulthood. The results from an earlier report (Robbins and Adams 1989) that did have an age breakdown therefore are given in Table 3.3. The Marshall

Islands data are of limited value in assessing the effects of I-131 exposure because more than 80 percent of the dose was from short-lived radioiodines and gamma radiation on these close-in atolls (Lessard and others 1985) rather than from I-131, and there were large uncertainties in the doses. In addition, many of the exposed subjects were put on thyroid suppression therapy beginning in 1965, but with variable compliance.

In 1987, a study was reported assessing whether there was an negative association between thyroid nodule prevalence on various Marshall Islands and the Bikini Atoll where shot BRAVO was detonated (Hamilton and others 1987). Residents on 14 atolls (n = 2,273) were given a thyroid screening, and information was obtained on where they lived at the time of the BRAVO shot. The researchers found a statistically significant association between distance and thyroid nodule prevalence, with additional accuracy of prediction if they added an assumed fallout cloud vector that initially went east, then southeast. A strength of the study is that the subjects had a second thyroid examination by a "blinded" examiner, and there was a good coefficient of agreement between the two examiners' results (kappa = 0.80). Weaknesses of the study include its reliance on palpation, the lack of dose information (risk estimates could not be computed), and the fact that the significant results were appreciably driven by the high nodule rates on the Rongelap and Utirik atolls, where much more thyroid screening occurred (nodules found in the past were included).

The prevalence of thyroid nodules has been investigated among residents of the island of Ebeye (Kwajalein Atoll); the island contains about one-quarter of the Marshall Islands population with former residents of many atolls now living there. Takahashi and others (1997) used palpation and ultrasound to examine 815 persons living in the Marshall Islands at the time of the BRAVO shot. Another 247 born after BRAVO but before the Bikini tests ended in 1958 and 260 born after 1958 also were examined. The researchers ascertained where each person in the first group resided at the time of the BRAVO shot. A marginal association (*p* = 0.08) was found for distance from Bikini and the prevalence of palpable thyroid nodules for the atolls that Hamilton and colleagues (Hamilton and others 1987) had used, and a similar association (*p* = 0.06) was found using all the atolls in their study. The results were slightly weaker for all nodular goiter (which apparently meant all nodules >2 mm in diameter detectable either by palpation or by ultrasound), with *p*-values of 0.07 and 0.12, respectively. Because there was thought to be reasonable similarity in the geographic distribution of cesium-137 and I-131, they correlated the mean Cs-137 measurements for the atolls in their study with thyroid nodule prevalence rates for those atolls. There was no association, calling into question the meaning of the Hamilton and others (1987) findings (although a limiting factor in the interpretation is the relatively low statistical power of this study because of the modest sample size). In addition, Takahashi and others (1997) found little association between the Cs-137 measurements and distance from Bikini, probably indicating that wind patterns, rainouts, and other factors were important influences on dose. More information from that study will be forthcoming; an additional 2,000+ Marshall Islanders have been examined and data analysis is in progress (Trott and others 1998).

In Utah, there have been two rounds of thyroid examinations (1965-1967 and 1985-1986) of a cohort of schoolchildren exposed to radioactive fallout, including I-131, from the nuclear-bomb testing in Nevada (Kerber and others 1993). The first round of screening included 4,818 schoolchildren. After excluding those who lived in states other than Utah, Nevada, or Arizona in 1985-1986, those with nonwhite or Hispanic ethnicity, and those with a history of radiotherapy, 3,180 of the original group remained of those screened in the first round, 2,473 were examined in the second round. The mean thyroid dose for the cohort was estimated at 0.098 Gy (9.8 rad) (Stevens and others 1992); the mean for the Utah subset of the cohort was 0.17 Gy (17 rad). It could be worth noting that the Utah study applied modeling formalisms and parameter definitions similar to that of the NCI (NCI 1997a) study, except to the assessment of individual doses. A complete description of the dose reconstruction methods for the Utah study can be found in Simon and others (1990).

The composite of the two rounds of screening in Utah plus any interim thyroid diagnoses showed 8 thyroid cancers in the irradiated group; about 5.44 would have been expected (Kerber and others 1993). The dose-response analysis was not statistically significant, but the central risk estimate (ERR, 7.9 per Gy) was similar to the pooled estimate derived from studies of external radiation. The strengths of the study include its careful effort at dose reconstruction, the use of information on childhood milk consumption, and the sophisticated statistical analyses that incorporate the joint prevalence-incidence data and the uncertainties in dosimetry. Its limitations include the small number of thyroid cancers detected with the consequent low precision of the results. Palpation was used as ultrasonography was not routinely available for screening. The screening examiners and the physician who ordered further diagnostic tests were not totally blinded in that they knew whether the subject was from a high or low exposure area, though they did not know the subject's dose status.

Data from Studies of the Chernobyl Accident

Results from studies prior to the 1986 nuclear reactor accident in Chernobyl were suggestive of a link between I-131 exposure and thyroid cancer but studies following the accident are generally regarded as conclusive. Reports from Ukraine, Belarus, and, to a lesser extent, Russia indicate a substantial excess of thyroid cancer among children exposed to I-131 fallout from the Chernobyl nuclear reactor accident in 1986. The increased incidence of childhood cancers in the Chernobyl region began 4 years after the incident, which released very large amounts of I-131, other short-lived radioiodines, and other radioisotopes including cesium, xenon, krypton, and strontium (Becker and others 1996). The youngest children diagnosed with thyroid cancer were only 4 years old at the time of diagnosis and were in utero at the time of the accident. The peak occurrence of thyroid cancer between 1990 and 1994 was for 8-year-old children in Ukraine and 9-year-old children in Belarus (Becker and others 1996). This is considerably younger than the usual age of onset (14 or older) for naturally occurring thyroid cancer elsewhere in Europe and the United States during the same period (Pacini and others 1997; Williams and Tronko 1996; Zimmerman and others 1994). Normally, thyroid cancer in young adults occurs predominantly in females, but after the Chernobyl accident the female-to-male ratios were below 2 (Pacini and others 1997; Becker and others 1996; Williams and Tronko 1996; Fuscoe and others 1992).

When the reports first began to appear about Chernobyl (Kazakov and others 1992), there was considerable skepticism that the cases represented a real radiation-induced increase. Five reasons for skepticism were cited including

- Suspicions that additional cases were a function of additional screening and surveillance, a common problem.
- Belief that I-131 was only weakly carcinogenic compared with externally generated x-rays.
- Inconsistency with past epidemiological studies with the Chernobyl cases occurring earlier after exposure.
- Questions about the accuracy of the thyroid dose estimates.
- Concerns that thyroid cancer incidence could have been enhanced because some regions were borderline goitrogenic because of low iodine concentrations in drinking water and foods.

More recent studies of the aftermath of Chernobyl have addressed these sources of skepticism. Questions about dosimetry remain, but a remarkably coherent picture of cancer risk related to I-131 exposure has emerged from the Chernobyl studies.

Screening and Surveillance Effect One argument against a substantial surveillance effect is that the proportion of tumors of stage T4 at diagnosis (the tumor had broken through the thyorid capsule and invaded surrounding tissue) was unusually high, some 40 percent of cases in Belarus (Kazakov and others 1992). Diagnosis at such a late stage in the development of the tumor is consistent with the claim that most cases were ascertained as a result of a visit by the child to a doctor and not through screening programs. In addition, studies also show a very marked decline in incidence rates to preaccident levels in children born after the end of 1986, even though screening is still common (Stsjazhko and others 1995). This indicates that the increase seen in those who were young at the time of the accident is related to some short-lived initiation process operative at the time of the accident, which is consistent with the increase being caused by exposure to the isotopes of iodine. In addition, because Belarus is an iodine-deficient area with a marked prevalence of goiter (Gembicki and others 1997; Nikiforova and others 1996), a system of surveillance for thyroid abnormalities was in place in schools at the time of the accident. This consisted of regular visits to schools by an endocrinologist to conduct palpation of the neck. There is no evidence that screening with ultrasound was widespread until 1991-1992. According to Williams and others (1996), 13,000 children had been screened in the Gomel region of Belarus by the end of 1994. Between 1986 and the end of 1994, 178 cases of childhood thyroid cancer were diagnosed in the same region. It is, therefore, clear that screening, even at that late date, was not the primary means of case ascertainment.

Relative Carcinogenic Effect of I-131 Among the reasons posited for the smaller carcinogenic potential of I-131, compared with externally generated x-rays, was the absence of an excess of thyroid cancer in the populations studied after the diagnostic administration of I-131. The results of a pooled analysis of 5 epidemiologic studies on infants, children, and adolescents irradiated with externally generated x-rays (Ron and others 1995), show that there is a steep decline in sensitivity, as measured by ERR, to cancer induction by x-rays with age at exposure. It was already known from the survivors of the atomic bombings in Japan that young adults were relatively insensitive and those over the age of 40 at small risk of thyroid cancer induced by the external gamma-rays from the bomb. Because I-131 is rarely given to children and only occasionally to adolescents, the lack of thyroid cancer associated with diagnostic administration of I-131 can be explained by the small number of children in the surveys and the insensitivity of the adult population (Ron 1996). As will be shown below, it is in fact the data accruing from the follow-up to the Chernobyl accident that are most likely to resolve this long-standing question.

Timing of Cases It has been claimed that too many cases occurred after Chernobyl too soon after exposure. In the pooled analysis of 5 non-Chernobyl studies cited earlier (Ron and others 1995) 2 cases in 81,000 person-years of follow-up were diagnosed within 5 years of exposure. In the Gomel region of Belarus 15 cases were diagnosed in the first 5 years in about 360,000 exposed (EP Demidchik, personal communication). The rates of appearance are therefore 25 and 8 per 1 million person-years, respectively, in the pooled studies and from Chernobyl. In view of the small numbers in the pooled study and the nonidentical dose distributions, there probably is no significance in the factor-of-3 difference.

Dosimetry Several measurements of activity in the thyroid were made in some settlements, but these were usually single measurements from which doses had to be derived using an assumed retention function. Attempts to relate I-131 activity to that of other longer lived nuclides, such as Cs-137, have not been successful, probably because of the differential release of the nuclides over the 10 days of the incident. The authenticity of such dosimetric estimates as there are can best be judged by the coherence of the estimates of risk.

In an analysis of Chernobyl thyroid cancer risk that was more quantitative than earlier studies (Jacob and others 1998), average thyroid doses were estimated for children in nearly 6,000 settlements in Ukraine, Belarus, and Bryansk, Russia. The authors estimated that, at least in Ukraine, the 95 percent CI around the imputed settlement childhood thyroid doses was about a factor of 2 above and below the estimated doses. The mean settlement doses were then used to derive average thyroid doses for children in various regions, which in turn were regressed on thyroid cancer rates. The analysts reported a good fit to a linear dose-response association, with a slope indicating an EAR of 2.3 (95 percent CI, 1.4-3.8) per 10^4 person-year-Gy. The EAR is about half as large as that reported in the pooled analysis of thyroid cancer after childhood external irradiation (Ron and others 1995), but both estimates are statistically compatible. Nevertheless, if one considers that there may well be at least a modest surveillance effect, then the difference would be somewhat larger if numerical account could be taken of the fact that this population with high surveillance is being compared with populations from several external radiation studies in which there was little special surveillance. Because background thyroid cancer rates are very low for young people, the ERRs in this study were very high (22-90 per Gy) and are probably not a good basis for projecting risk as the group ages.

Measurements of iodine activity in the thyroid made in the few months after the accident in Belarus indicate doses ranging up to several Gy (OECD 1996). As in the case of the weapons testing in Nevada, the dominant contribution to dose from radioiodine is from I-131 transmitted from ground deposition on pasture through the food chain in milk. The shorter lived isotopes of iodine, relatively less abundant in fallout from a nuclear reactor than from an atomic weapon, are estimated to contribute less than 15 percent to the thyroid dose for those exposed after Chernobyl. I-131 thus seems overwhelmingly implicated in the cases arising after the Chernobyl accident.

A survey of activity in the thyroids of children living in the Gomel region of Belarus indicated that, for children aged 7 years or younger, the average absorbed dose was about 1 Gy (100 rad) (OECD 1996). About 162,000 such children were resident in Gomel at the time of the accident. The risk factor derived from the pooled study (Ron and others 1995) is 4.4 cases per 10^4 person-year-Gy. For 162,000 such children exposed to 1 Gy there should be about 71 cases per year averaged over the some 40 or more years during which the expression of the cancer takes place. Since the accident and up to 1995, there have been 122 cases in children aged 7 and younger, the greater proportion occurring since 1991. Thus, the average rate over this period is about half that expected in the longer term. This crude estimate is consistent with a recently published study (Jacob and others 1998) of children under 15 at the time of the accident in settlements in the affected countries. The pooled data EAR is 2.3 (95 percent CI, 1.4-3.8) per 10^4 person-year Gy. This point estimate is about half the point estimate for the 5 pooled studies of children exposed to external radiation (Ron and others 1995). However, this estimate is influenced by a study from Israel; were this study excluded, the point estimates would be nearly coincident. In any event, the estimate from Chernobyl lies within the 95 percent CI for the x-ray studies (Ron and others 1995), which indicate a peak in incidence rate some 15-19 years after exposure. Estimates for thyroid cancer among the Chernobyl population could be expected to increase, rather than decrease, with time.

The concordance between what is seen after Chernobyl and what is understood from epidemiologic studies of children treated with x rays is confirmed by another approach to determining the relationship between dose and effect. By comparing the cumulative incidence of thyroid cancer in those under the age of 15 at the time of the accident with the cumulative incidence in children irradiated for enlarged thymus gland in the United States (Shore and others 1993), it can be deduced that children in the Gomel region received from I-131 the equivalent of about 1 Gy (100 rad) of externally generated x rays. The average dose to children under the age of 8 is reported to be 1 Gy (OECD 1996), indicating an average dose to children under 15 of about 0.8 Gy (80 rad). Allowing for the early stage in the expected evolution of solid cancers after exposure to radiation, there is little room for a very markedly reduced efficiency of cancer induction by I-131 compared with that by externally generated x rays.

Contribution of Dietary and Other Factors Finally, the question arises whether the effects of radiation have combined with factors such as iodine deficiency and lifestyle to increase the extent to which cancer has been observed in the countries adjacent to Chernobyl. Dietary iodine deficiency is evident in the regions surrounding Chernobyl (Nikiforova and others 1996). It could increase the effectiveness of I-131 both by increasing the uptake of radioactive iodine and by accelerating the appearance of initiated disease through the increased expression of thyroid-stimulating hormone (TSH). The first of these effects seems unlikely to be important when iodine is taken into the body in small quantities over a prolonged period, as is the case with environmental exposure. Although Nikiforova and others (1996) failed to find a significant correlation between urinary iodine concentration and TSH concentration in some 5,000 children they studied, elevated concentrations of TSH have been associated with low urinary iodine and goiter prevalence.

About half of the people who live near Chernobyl lead a rural lifestyle and get their milk from private or backyard cows. The short cow-to-consumer interval and the increased possibility of being exposed to the short-lived isotopes of iodine and tellurium-132, the precursor of I-132, may increase the dose relative to people who live in towns. There is, however, no evidence that either of these factors has had a dominant effect on the expression of thyroid cancer after Chernobyl.

Insofar as can be deduced from the early stages of the increase in thyroid cancer observed after Chernobyl, the risk is not less that 50 percent of the risk attributed to x-ray exposure of the child thyroid and, because of the early stage, could well prove equal. Thus, of the values commonly cited for relative biological effectiveness (RBE) (0.1, 0.33, 0.67, 1.0) or the dose and dose-rate effectiveness factor (DDREF) (10, 3, 1.5, 1.0), for irradiation of the thyroid by I-131, only the latter two (RBE, 0.67 or 1.0; DDREF, 1.5 or 1.0) receive support from the Chernobyl experience.

Histologic and Biologic Features of Thyroid Cancer in Children of Chernobyl Exposed to Radiation from Radioiodine

Most of the thyroid cancers (96-99 percent) that have occurred in children exposed to radioactive iodine from the Chernobyl accident are papillary thyroid cancers, but they seem to be more aggressive than usual. However, naturally occurring childhood thyroid cancers are also more aggressive, but not clearly more lethal, than those in adults (Pacini and others 1997; Becker and others 1996; Nikiforov and Gnepp 1994). There are certain histologic variants of papillary thyroid cancer that seem to occur with higher than usual frequency in the children of Chernobyl exposed to I-131 (Williams and Tronko 1996). In contrast, the histology of papillary cancers that occur after exposure to external radiation does not differ much from those that occur spontaneously, and the long-term outcome of the two is similar (see, for example, Viswanathan and others 1994).

Microscopically, some of the tumors in Chernobyl-exposed children seem to be especially aggressive, as evidenced by intraglandular tumor dissemination (92 percent), thyroid capsular and adjacent soft-tissue invasion (89 percent), and cervical lymph node metastases (88 percent) that occurred at a greater rate than usual. These observations, however, were not blinded as to the source of the tumor, and some bias could have been introduced (Nikiforov and Gnepp 1994). Another study compared post-Chernobyl thyroid cancers in Belarus children and adolescents with those occurring naturally among children in Italy and France (Pacini and others 1997). In the Belarus cases, extrathyroidal extension (49.1 percent, $p = 0.001$) and lymph node metastases (64.6 percent, $p = 0.002$) were more frequent than in the cases from Italy and France (24.9 percent and 53.9 percent, respectively). But distant metastases, potentially the more dangerous, were found in 7.8 percent of the Belarus group and in 17.3 percent of the Italy-France group.

The disease in children exposed as a result of the Chernobyl accident appears to have a short latency period, a higher proportion of tumors arising in young children (under age 5 to 8 years), and an almost equal sex ratio. Both young age at disease onset, which could merely be a reflection of the short follow-up and the fact that young children are most sensitive to the effects of radiation, and radiation etiology appear to increase the aggressive growth of thyroid cancer (Pacini and others 1997; Robbins 1994). It is too soon to determine whether the long-term prognosis for survival in children exposed to radiation during the Chernobyl accident will be different from that for spontaneously occurring papillary cancer, although there are early indications these cancers are more aggressive. Future studies will indicate whether they are more life threatening.

At the molecular level, the radiation-induced tumors appear to be different from those that occur spontaneously. One study (Ito and others 1994) found *ret* activation in 57 percent (4/7) of thyroid cancer cases from Chernobyl. Another (Fugazzola and others 1995) found it in 66 percent (4/6), and a third study (Klugbauer and others 1995) reported it in 64 percent (9/14) of the cases from Chernobyl. Although there is an increased frequency of PTC rearrangements in young patients and in children unexposed to thyroid irradiation who develop thyroid cancer (Bongarzone and others 1996), it appears that *ret* activation is more frequent in the Chernobyl cases.

Moreover, chromosome rearrangements forming PTC3 and novel *ret*/PTC rearrangements seem to be more frequent in the cases from the Chernobyl accident. A larger controlled study (Nikiforov and others 1997) found 76 percent of 38 thyroid cancer cases from Belarus had *ret* rearrangements, 6 to form PTC1, 1 to form PTC2, and 22 to form PTC3. In comparison, 65 percent of 17 unexposed cases had *ret* rearrangements. Two variants of PTC3 also have been found. One variant (Klugbauer and others 1995) is a rearrangement lacking one exon of the ELE1 gene. The other variant (Fugazzola and others 1996) is a rearrangement with an additional 93 base pairs derived from the *ret* gene. A fifth novel *ret* rearrangement, PTC5, was detected in papillary thyroid cancers of 2 patients exposed to radioactive fallout after Chernobyl (Klugbauer and others 1998). PTC5 features a fusion of the *ret* tyrosine kinase domain with a sequence identical to that previously described as *ret2*, which is a transfection artifact in NIH3T3 cells not previously detected in human tumors. The ubiquitous *ret*-fused gene 5 was found in various normal tissues, including the thyroid gland.

NONMALIGNANT THYROID DISEASE ASSOCIATED WITH RADIOIODINE EXPOSURE

Data on the induction of nonmalignant thyroid disease are inconclusive in the I-131 dose range to which most people were exposed from Nevada Test Site fallout. Additional, possibly more conclusive information on the link between low to moderate doses of I-131 and thyroid diseases other than cancer should soon be available from the Hanford Thyroid Disease Study.

The thyroid is unusually susceptible to radiation exposure through its ability to concentrate radioiodine and because of its anatomic position, which is commonly in the field of external radiation applied for therapeutic purposes. Acute radiation doses that cause potential cell cycle disturbances (short of immediate cell death) do not cause obvious injury in the short term until the irradiated cells attempt to divide—then they experience mitotic death. The thyroid follicular epithelium has a long turnover time (1-2 years), so evidence of direct radiation injury can be delayed. The degree of impairment of reproductive capability of the thyroid follicular epithelium is related to the radiation dose, with a possible lower threshold between 2 Gy (200 rad) and 4 Gy (400 rad) for this acute injury (Williams 1991). Beyond this, the pituitary axis can compensate for a considerable degree of thyroid injury with increases in TSH production.

It is well documented that I-131 therapy for thyrotoxicosis at radiation doses of 50 Gy (5,000 rad) and above is associated with destruction of the thyroid gland and hypothyroidism (Maxon and Saenger 1996; Cooper 1991). This treatment initially causes an inflammatory response, followed by long-term chronic inflammation, fibrosis, and atrophy, ultimately resulting in thyroid gland failure. According to Cooper (1991), hypothyroidism may be considered an inevitable consequence of radioiodine therapy. Up to 90 percent of patients are affected in the first year after therapy, with a continuing rate of 2-3 percent per year thereafter. The occurrence of hypothyroidism relates to radiation dose, but also to prior status of antithyroid antibodies in irradiated patients, there being a correlation between the occurrence of thyroid autoantibodies both before and after treatment and development of hypothyroidism (Cooper 1991; Lundell and Holm 1980).

High radiation doses clearly can cause direct follicular cell injury, but there is considerable evidence that this is not the only factor involved in radiation-related nonneoplastic thyroid disease. Various forms of primary thyroiditis are considered to have an autoimmune pathogenesis. Both hypothyroidism (including Hashimoto's and atrophic thyroiditis) and some forms of hyperthyroidism (Graves's disease) are caused by closely related genetic and immunologic disturbances, although they are considered distinct diseases with separate defects in immunoregulation. Autoimmune thyroiditis leading to hypothyroidism is related to the presence of autoantibodies directed against some component of the thyroid gland, such as thyroglobulin or thyroid cell microsomes, the latter being considered more significant (Volpé 1991a). In Graves's disease, there is diffusely hyperplastic goiter resulting from production of antibodies directed against the TSH receptors, which cause excessive stimulation of the follicular cells (Volpé 1991b). It is well known that both hypothyroidism and hyperthyroidism associated with autoimmune disease can be induced by exposure to ionizing radiation (DeGroot 1988).

After I-131 therapy for hyperthyroidism, the occurrence of hypothyroidism depends on the dose, and as mentioned previously, on the nature of the autoimmune response. There is evidence of persistent thyroid-stimulating antibodies and other autoantibodies for several years after I-131 treatment in some patients (Williams 1991). Hypothyroidism also can occur after external exposure from radiotherapy when the thyroid is in the field, for example, in patients treated for lymphomas, lymphoblastic leukemia, and various head and neck cancers (Tell and others 1997; Hancock and others 1991; Williams 1991; Fleming and others 1985). In a group of patients with prior irradiation for benign head and neck disorders, some sort of "radiation thyroiditis" was found in resected thyroids from 10 to 75 years after exposure (Swelstad and others 1977). The foregoing cases were associated with therapeutic doses of radiation generally measured in the tens of Gy, but there is also ample evidence for hypothyroidism induced by somewhat lower doses. Thyroid hypofunction was reported in fallout-exposed Marshall Islanders, starting about 10 years after exposure and was most marked in those who were exposed as children under the age of 6 (Larsen and others 1982). Estimated doses to persons showing thyroid hypofunction ranged from 1.35 to 21 Gy (135 to 2,100 rad). Shorter-lived iodine isotopes (I-132, I-133, I-135) were responsible for a significant component of the dose in this population.

Data from the atomic bomb survivors are conflicting. Nagataki and others (1994) reported a significant, linear-quadratic, dose-response relationship for autoimmune hypothyroidism in 17.3 percent of 2,587 exposed subjects in Nagasaki. The response peaked at about 0.7 Sv (70 rem), but had an apparent increase in the 0-0.5 Sv (0-50 rem) dose range. It was also noted that the prevalence increased with each passing decade after exposure. Another study of atomic bomb survivors in Hiroshima reported a higher rate of hypothyroidism in both males and females in the 0.01 Gy (1 rad) to 0.99 Gy (99 rad) cohort, although paradoxically the antithyroid microsomal antibodies were decreased in this irradiated population (Ito and others 1987). Morimoto and others (1987) reported no association between radiation and hypothyroidism in Hiroshima and Nagasaki survivors who were less than 20 years old at exposure. Yoshimoto and others (1995) reported no statistical association between radiation exposure and chronic thyroiditis in Hiroshima atomic bomb survivors. Fujiwara and others (1994) found no effect of radiation on the prevalence of antithyroglobulin or antithyroid microsomal antibodies in atomic bomb survivors in Hiroshima and Nagasaki.

Reports on the populations exposed to radioiodine after the Chernobyl accident are also of interest. Children exposed to I-131 had increased autoantibodies to thyroglobulin, thyroid hormone, and TSH; antithyroid antibodies were increased in a dose-responsive fashion, including apparent increases at doses below 1 Gy (100 rad) (Vykhovanets and others 1997). It has been suggested that irradiation is associated with increased risk for chronic thyroiditis (Ito and others 1995). The Chernobyl effects are complicated by the presence of iodine deficiency in some affected areas, although Kasatkina and others (1997) report increased antithyroglobulin and antithyroid microsomal antibodies in children with both radiation exposure and poor iodine intake, suggesting a possible interaction.

A study of 297 children under the age of 16 exposed to diagnostic I-131 at an average dose of 3.8 Gy (380 rad) showed an increased risk for developing hypothyroidism up to 14 years after exposure; however, there were no cases of hypothyroidism in 146 children with a mean thyroid dose of about 0.18 Gy (18 rad) (Maxon and Saenger 1996).

Other studies of relatively low dose irradiation have not confirmed increased nonmalignant thyroid disease risk. Thymic irradiation during infancy resulted in no observable thyroid disease, hormonal abnormalities, or antithyroid antibodies at exposures of 30-1,200 roentgen (R) x rays (Hildreth and others 1987). Irradiation for childhood hemangiomas with a variety of methods, including α, β, γ, and x-irradiation, resulted in no thyroid functional abnormalities at doses ranging from <0.01 to 2.74 Gy (1-274 rad), although thyroid nodules were increased (De Vathaire and others 1993). A study of the Utah "downwinders" exposed to atomic weapons test fallout after birth (mean dose of 0.098 Gy [9.8 rad] with a maximum of 4.6 Gy [460 rad]) showed no excess of nonmalignant thyroid disease (Kerber and others 1993).

There are reports of hyperthyroidism after irradiation of the thyroid gland for different conditions, including breast cancer (Williams 1991), gynecologic cancers (Katayama and others 1985), and Hodgkin's disease (Hancock and others 1991; Loeffler and others 1988). Hyperthyroidism after treatment of goiter with I-131 was associated with increased serum concentrations of TSH receptor antibodies (Huysmans and others 1997). It has been suggested that Graves's disease induced by radioiodine is associated with a release of antigen from the damaged thyroid and the subsequent production of antibodies that stimulate the TSH receptors (Kay and others 1987). It also has been suggested that suppressor lymphocytes in the thyroid might be more sensitive to radiation than are helper cells, increasing the autoimmune response (Williams 1991).

Overall, these data clearly indicate that there is a highly significant association between ionizing radiation exposure and the occurrence of nonmalignant thyroid disease at higher levels of exposure. It also is evident that this risk could extend down into the range of doses below 1 Gy (100 rad). For exposure to I-131, however, Maxon and Saenger (1996) indicate that hypothyroidism from I-131 would be unlikely at doses below 0.1-0.2 Gy (10 to 20 rad).

Thus, the data on nonmalignant disease induction are inconclusive in the dose range to which most people were exposed from fallout. As a result, the IOM/NRC panel did not further consider the implications of nonmaligant disease. However, the current Hanford thyroid disease study is evaluating all thyroid diseases, including autoimmune thyroiditis and hypothyroidism and antithyroid antibody concentration, in association with radiation dose. The data from this study will be extremely valuable in analyzing risk for nonmalignant thyroid disease. Depending on the results, NCI should consider how to incorporate such information in any program to communicate about the risks from fallout from the Nevada nuclear weapons test and whether to undertake a formal evaluation of the screening for nonmalignant disease.

THYROID CANCER RISK BASED ON NCI ESTIMATES OF
I-131 DOSES

The 1997 NCI report did not attempt to translate the thyroid dose estimates that it derived into risk of thyroid cancer or other thyroid diseases. However, simultaneous with the release of the report, a staff memorandum prepared by Dr. Charles Land presented calculations of lifetime thyroid cancer risk and corrected calculations were presented to the NRC committee in December 1997. (Appendix B presents this memo.) The corrected calculations yield an estimated 11,300 to 212,000 excess cases of thyroid cancer resulting from the Nevada weapons tests.

The Land estimates of lifetime thyroid cancer risk rely on a linear model for ERR of thyroid cancer as a function of two factors: thyroid dose and age at exposure. The statistical model for thyroid cancer (as well as the dose-response estimates used in the calculations) relies on meta-analysis of thyroid cancer caused by external radiation of the thyroid that was discussed earlier in this chapter (Ron and others 1995). Allowance for the possibly reduced RBE of beta-rays from internal exposure to I-131 compared with external doses also was considered in Land's calculations.

The basic features of the model used by Land to derive excess cancer estimates are (1) linearity (even the smallest dose to the thyroid results in some ERR for cancer), (2) additivity of the effects of multiple exposures (over the period of the Nevada atomic weapons tests) on risk, and (3) an assumption that irradiation of the thyroid in childhood leads to an elevated relative risk of thyroid cancer for a person's lifetime. These assumptions together mean that calculations of the excess fraction of thyroid cancer cases for a given age group caused by exposure are given by computing average doses for persons exposed at that age and multiplying the averages by age-specific excess risk estimates.

Table 3.4 gives ERR values per unit dose, according to age at exposure, from Ron and others (1995), and the estimated average dose for each age group from the NCI (1997a) report. The relative risk for thyroid cancer for a person exposed at that age, compared with an unexposed person of the same age and gender, is calculated using an RBE of 0.66, consistent with the concluding estimate of the Land analysis (which also presented estimates based on other RBE values). The committee believes that an RBE of 1.0 would be equally acceptable given the scientific information available at this time.

Table 3.4 also gives lifetime thyroid cancer risk for males and females for each age-at-exposure group and total excess cases within each age group. The lifetime risks are obtained by multiplying the SEER 1972-1992 baseline risk estimates (0.25 percent for males, 0.64 percent females, which are rates over 85 years for all forms of thyroid cancer including those not linked to radiation exposure) by the relative risk for each age-at-exposure group. According to this table, for a woman exposed to 0.1 Gy (10 rads) of I-131 at less than one year of age, the

TABLE 3.4 ERR per Unit Dose, Relative Risk, and Excess Cancer Cases, by Age at Exposure

Age at Exposure	ERR at 1 rad[a]	Average Dose (rad)	Relative Risk[b]	Percentage Lifetime Risk Male	Percentage Lifetime Risk Female	Excess Cancer Cases
<1	0.098	10.3	1.67	0.42	1.07	10,170
1-4	0.098	6.7	1.43	0.36	0.92	26,993
5-9	0.049	4.5	1.15	0.29	0.73	9,064
10-14	0.0245	2.8	1.05	0.26	0.67	2,451
15-19	0.01255	1.8	1.01	0.25	0.65	748
>20	Negligible	1.8	1.00	0.25	0.65	0

[a] ERR values are taken from Ron and others (1995).
[b] Relative risks are computed using an RBE of 0.66 consistent with the analysis of Land (see Appendix B).

lifetime risk to age 85 of being diagnosed with thyroid cancer is estimated to increase from 0.64 percent to 1.07 percent, which would still make a diagnosis of thyroid cancer uncommon compared to a number of other cancers.

The period of aboveground testing in Nevada extended from 1951 to 1958, so exposure to I-131 would have occurred over a range of ages for most persons. However, the combination of the linearity and additivity assumptions implies that doses given at specific ages to different people can be combined in calculating excess cancer cases. An infant in 1951 moves to the 5-9-year category by 1958, but is replaced by other infants born in the intervening years, with these two effects essentially canceling out in the computation of excess cases for the <1-year age category. Thus, Land uses the U.S. population age distribution in 1952 and assumes that all exposures occurred in 1952 to approximate excess cases resulting from exposures that were actually distributed over a longer period to a larger number of people.

The Land analysis notes that there are substantial uncertainties in the estimate of ERR at 1 rad and the average dose for each age group. An uncertainty analysis, which allowed for the 95 percent confidence interval (0.021-0.287) reported by Ron and colleagues (1995) for the ERR estimate at 1 rad and a factor of 2 in the uncertainty of average thyroid dose, gave a range for the total excess cases due to exposure of 11,300-212,000. The central estimate implied (but not explicitly stated in that analysis) is 49,000 cases. An independent estimate of the uncertainty in the number of excess cases produced a 95 percent confidence range from 8,000 to 208,000. This latter analysis also included the uncertainty of the RBE for I-131 (F.O. Hoffman, communication to NAS, 20 December 1997). Beyond this uncertainty assessment, the validity of the assumptions of linearity, constancy of ERR, and the additivity of effects of exposure are all critical to Land's estimate of excess cases. Both linearity and extended elevated risk are consistent with the publications reviewed by Ron and colleagues (1995). However, these data do not provide a clear test of these assumptions at either the range of doses given by NCI (1997a) or over the ranges of time since exposure that are important today. The highest average dose considered in Land's calculations (for infants) is just above the average thyroid dose (9 rad or 0.09 Gy) reported to have an effect on risk in one study. Although excess cases of thyroid cancer were found by the studies analyzed by Ron and colleagues (1995) to continue for long periods after exposure (up to 40 years) with little falling off in ERR, no study has followed exposed subjects over an entire lifetime. Today, the most heavily exposed subjects to I-131 fallout from the Nevada Test Site are nearly 45 years beyond their time of exposure, which is at the outer limits of follow-up of the subjects in the epidemiologic surveys.

Analyses of the thyroid cancer cohorts and case-control studies have shown important heterogeneity in the dose response for different groups of subjects. In reflecting this heterogeneity, Ron and colleagues (1995) were led to use a random-effects model when combining data from the 7 studies analyzed to form an overall estimate of ERR per unit dose. Only one of the studies supports a thyroid cancer effect at dose ranges consistent with average thyroid dose to large birth cohorts from Nevada weapons tests. This study of Israeli children treated with radiation for *tinea capitis* showed a much stronger dose response than was evidenced in other studies. Although causes of heterogeneous results between studies are difficult to determine, Shore has suggested (1992) that the subjects in the Israel study might have had a higher dose response because of a higher-than-normal prevalence of inherited *ataxia telangiectasia* (A-T) heterozygosity in the population. This disorder is associated with a defect in DNA repair that could increase sensitivity to cancers of many types in A-T patients and could do the same in the heterozygote. If this study were dropped from the analysis, the estimates in the Land memo would change.

For setting prospective dose limits for radiation protection, the assumption of linearity and the use of linear extrapolation is widespread. For example, in the absence of direct data on health effects at low levels of radiation exposure, major regulatory affiliated organizations (International Commission on Radiological Protection, National Commission on Radiological Protection, Environmental Protection Agency, National Research Council, and others) extrapolate all radiation effects through the zero effect-zero dose point in a linear fashion.

Although opinion leans heavily toward use of linear no-threshold methods of estimating risk for thyroid disease, this concept is *not universally accepted*. It clearly introduces another uncertainty factor into the calculation of risk. If the dose response were nonlinear, that could produce a lesser or greater response at low doses, and would reduce or increase the overall risk of health effects from exposure to I-131.

Uncertainty also applies both to the assumption of constancy of ERR and to the assumption of additivity of multiple exposures. This committee notes that natural background doses over the 10-year period of testing would present to each person in the United States an average dose that was approximately half of the average dose from the weapons testing (2 rad or 0.02 Gy). The amount of natural background radiation varies geographically, such that, in some areas, the 10-year average could have been lower or higher by 0.01 Gy (1 rad), when radon exposures are ignored, bringing the average dose to 0.03 Gy (3 rad) or 0.04 Gy (4 rad) to the thyroid over the 10 years of testing. Some areas would have been lower and some higher in the calculated fallout dose ranges. It would be informative, therefore, to determine how background radiation would alter total dose, and hence risk, to persons over this 10-year period. Calculations could be done to compare the total doses using the actual background dose for each county to add to the total estimated dose to the average individual in each county, but this would require substantial effort.

Analyses of the risk of thyroid cancer after exposure to ionizing radiation such as those analyzed by Ron and colleagues essentially assume similarity in naturally occurring and radiation related thyroid cancers. Although this chapter has cited histologic evidence suggesting that the differences in thyroid cancers in children exposed from the Chernobyl accident, there are no data yet relevant to effects in adults exposed as children to the I-131 levels associated with the Nevada weapons tests.

Given some of the issues noted above, the committee suggests that DHHS consider some additional analyses to evaluate further the estimated confidence intervals for the risk projections and to improve understanding of the sensitivity of the projections to changes in key assumptions. Such analyses might include (1) use of alternative dose-response models, (2) choice of different average population doses, (3) use of a model of excess relative risk that declines as a function of years since exposure, and (4) exclusion of the *tinea capitis* study from the Ron analysis.

As indicated above, there are few data to show a statistically significant carcinogenic effect of radiation to an organ or whole body below a dose of 0.1 Gy (10 rad). Epidemiologic studies, which might be helpful, are complicated when estimated doses are low. For example, very large samples are needed to demonstrate an effect. Results of some epidemiologic analyses of the possible effects of I-131 fallout from the Nevada weapons tests are discussed below.

Epidemiologic Analyses Using Cancer Registries

Neither the NCI (1997a) report nor the analyses provided by Land (Appendix B, this report) consider whether there is any epidemiologic evidence of increases in thyroid cancer from exposure to Nevada Test Site fallout. Data from various tumor registries around the country could be useful in reducing the uncertainty in the estimate of the collective excess of thyroid cancer cases. In particular, analysis of registry data allows for the comparison of thyroid cancer incidence between birth cohorts within the same general locality. The NRC panel had access to three such analyses—one undertaken by the committee, one for the state of Idaho, and one prepared by NCI analysts. In addition, it reviewed the Utah study (Kerber and others 1993) cited earlier in this chapter. That study, which was published in the *Journal of the American Medical Association*, found more thyroid cancers that would be expected for a nonexposed population and produced a risk coefficient of 7.9 per Gy, almost the same value as calculated by Ron (7.7 per Gy).

To get an idea of the power of epidemiologic studies, consider the use of tumor incidence data from registries reporting (as do the current SEER data) all thyroid cases occurring over the years 1973-1992. Note that, according to Land's analysis, there is a population of 17.6×10^6 who were 0-4 years old in 1952, and thus were 21-44 between 1973 and 1992. Persons who were 0-4 years old in 1963 were born during the era of underground testing, and thus may be considered to have been largely unexposed. Their attained ages over the same 1973-1992 period will range from 10 to 33 years. Thus, we may compare, for example, the rates of thyroid cancer incidence among persons aged 25-29 between these two birth cohorts.

The average age of the earlier birth cohort over the time of atmospheric testing (1951-1958) is approximately 3.5 years. From Table 3.4, this corresponds to an average thyroid dose of approximately 0.067 Gy (6.7 rad), which, in turn, corresponds to an estimated excess risk of thyroid cancer of approximately 1.43 for this cohort compared with the later, unexposed birth cohort.

Assuming that the rate of thyroid cancer in the unexposed cohort is 4.5 per 100,000 person-years (SEER data), to have sufficient statistical power (90 percent) to detect such an excess incidence, the monitoring system would have to have covered approximately 617,000 people in each of the birth cohorts. This would yield an expected 199 cases of thyroid cancer in the exposed cohort and 139 in the unexposed cohort. If we use a risk factor 4.3 times smaller than median value of 7.7 per Gy (which gives the lower end of Land's CI), then approximately 9.9 million persons would have had to have been monitored from each birth cohort to have the same power to detect a 10 percent increase. The SEER registries cover about 10 percent of the entire U.S. population, or about 1.8 million persons from the 1948-1952 birth cohort born. Potentially, then SEER registry data could help assess whether past incidence in these birth cohorts is consistent with the higher part of the estimated range of excess risk provided by Land.

For this report, the committee examined thyroid cancer incidence rates by birth cohort using SEER data. It also reviewed an analysis of Idaho registry data and another analysis of SEER data (Gilbert and others, in press for the *Journal of the National Cancer Institute*) that used incidence data from 1973-1994 and thyroid cancer mortality data from the 48 contiguous states from 1957 to 1994. The approach of Gilbert and colleagues was to relate county-level geographic variation in thyroid cancer incidence and mortality with the NCI's county-by-county estimates of thyroid dose. This committee took a different approach. It compared "exposed" versus "unexposed" birth cohorts for differences in thyroid cancer rates. The analyses of Gilbert and colleagues are performed within each birth-cohort separately, so that no specific comparison of "unexposed" versus "exposed" birth cohorts is given. The two approaches—variation by birth cohort in SEER and in Idaho, and variation by geographical region in SEER and in the mortality data—are complementary, but each is subject to certain limitations. For example, the Idaho analysis is limited because of the small population of the state, which, in turn, limits the numbers of thyroid cancers to relatively small numbers. The birth cohort approach to the SEER data ignores the geographical variation in estimated dose. Also, the interpretation of results may be complicated by external factors such as secular trends in thyroid incidence and by exposure to I-131 from other sources such as the Pacific or Siberian nuclear tests, which continued beyond the time of above ground testing at the NTS. The analysis of Gilbert and colleagues avoids the latter complexity, but introduces other, perhaps more important problems, because county level doses are known with far less certainty than are the average doses for the birth cohorts. Moreover, the analysis of Gilbert and colleagues is much more affected by the results of migration from county to county either at the time of exposure or in later life.

Publicly available thyroid cancer incidence data from 9 tumor registries that make up the SEER program are tabulated in Table 3.5 for the two birth cohorts described above, for the attained ages of 25-29 years, by region (far west and the rest of the country). The far-western registries are in San Francisco-Oakland, Hawaii, and Seattle; the other, not-far-western registries are in Connecticut, Detroit, Iowa, New Mexico, Utah, and Atlanta. The far-western areas were much less exposed to Nevada Test Site fallout than was the rest of the country, so their incidence data are tabulated separately, although considerable migration from eastern to western areas can be assumed to have taken place.

For the registries outside the far west, there is some evidence of about a 10 percent excess of thyroid cancers in the exposed birth cohort, but the estimate is not statistically significant. An evident excess for males in the far-western registries is unexpected, but the numbers of cases, 38 and 28, on which this observation is based are not large. Given the NCI dose estimates, such excess cases are not likely to be attributable to exposure to Nevada Test Site fallout.

Comparisons of thyroid cancer rates in an earlier birth cohort, 1938-1942, with rates among the 1948-1952 birth cohort also are of interest (Table 3.6). The 1938-1942 birth cohort was aged 9-20 during 1951-1958. According to Table 3.4, the estimated excess risk of thyroid cancer would have ranged from 5 to 15 percent for this cohort, compared with more than 40 percent for the 1948-1952 cohort. SEER data allow us to compare rates of cancer at ages 35-39 for these two cohorts (Table 3.6). Again, about a 10 percent excess risk for the 1948-1952 birth cohort is seen in the non-far-western registries, but this is not statistically significant. The data from the far-western registries are not consistent between males and females; males show 32 percent fewer thyroid cancer cases in the 1948-1952 cohort, and females show a 23 percent increase. Because differences in thyroid cancer incidence in the western registries are less apt to be due to I-131 exposure from the Nevada Test Site than are differences in the other registries, they suggest some difficulties in attributing differences in the non-far-western cohorts to I-131.

Appendix D provides a detailed examination of the thyroid cancer rates in Idaho by birth cohort and county for the period 1970-1996. While several Idaho counties showed elevated rates of thyroid cancer, they did not correlate very closely with estimated thyroid dose. The birth cohort born from 1948 to 1958 showed a 5 percent excess of incident cases compared to earlier and later cohorts, but this increase was not significant with the 95 percent confidence interval as the excess among the counties ranges from −7 to +18 percent. Within the most sensitive of the three birth cohorts examined in the Idaho analysis, there was little evidence of association with county or the county-specific estimate of thyroid dose; however the power to detect a dose response, within the relatively small population of Idaho, is quite low.

The analysis of Gilbert and colleagues relates geographic variation in thyroid cancer incidence and mortality rates in counties from the 48 contiguous states to the NCI estimates of average dose within each county. Overall that analysis finds negative nonsignificant excess risk due to I-131 exposure. However, when the analysis is restricted to those subjects exposed in infancy (age <1 y), a positive association between dose and risk is marginally significant for both thyroid cancer incidence ($p = 0.11$) and mortality ($p = 0.054$). The findings for that very narrow age group (<1 y at exposure), however, contrast with those in the group aged 1-5 years at exposure, for which dose responses were estimated to be negative despite the fact that subjects aged 1-5 years ought to be nearly as radiosensitive as the infants (based on findings from the external radiation studies). When the analyses of Gilbert and colleagues were restricted to subjects who were aged 0-4 at any time during 1950-1954 (i.e. those born between 1950-59), a significant association between I-131 exposure and thyroid cancer mortality (ERR per Gy = 12.0, p = 0.005) was detected. No other birth cohort (including those born between 1955 and 1964) showed positive risk estimates for mortality based on the county level dose estimates. Moreover there was no increase in incidence detected in the 1950-1959 birth cohort (ERR per Gy = 0.3, p = 0.66) The fact that mortality but not incidence showed a dose response in this birth cohort is puzzling. There were almost 3 times as many incident cases as thyroid cancer deaths (12,657 vs. 4,602), which would imply that the power to detect increases in risk should be far greater in the thyroid cancer incidence data than in the mortality data.

The impression left by these three studies—birth cohort comparisons of incidence in Idaho, birth cohort comparisons in SEER, and the study of Gilbert and colleagues using geographical variation in SEER incidence and thyroid cancer mortality—is that there is little evidence of widespread increases in risk related to

TABLE 3.5 SEER Thyroid Cancer Incidence Data. Comparisons of Thyroid Cancer Rates at Ages 25-29 Between Birth Cohorts Born in 1948-1952 (Exposed) and 1959-1963 (Relatively Unexposed)

Birth Cohort	Number of 25-29 Year Olds	Number of Thyroid Cancers	Approximate Rate per 100,000 Person-Years	Percentage Excess in Birth Cohort	p-Value for Excess
Far western registries					
1948-1952 (Monitored in 1977)					
Males	317,724	38	2.39	65.6	0.10
Females	310,404	136	8.76	8.1	0.50
1959-1963 (Monitored in 1988)					
Males	387,623	28	1.44	—	
Females	367,614	149	8.11	—	
Other registries					
1948-1952 (Monitored in 1977)					
Males	590,102	52	1.76	9.8	0.63
Females	601,506	257	8.55	9.9	0.28
1959-1963 (Monitored in 1988)					
Males	673,052	54	1.60	—	
Females	683,896	266	7.78	—	

the pattern of exposure of I-131 as described in the NCI report. They suggest that the numbers of excess cases of thyroid cancer due to I-131 exposure from the Nevada weapons tests are likely to be in the lower part of the range estimated in the Land memo. Furthermore, the estimate of only a 10 percent increase in risk between the exposed and nonexposed birth cohorts as detected in SEER, lies in the lower portion of the range of excess cases that was estimated by Land. Translating the 10 percent increase in risk for these cohorts to excess lifetime risk (using the linear constant excess relative risk model) amounts to approximately 11,300 excess cases (using an RBE of 0.6). The range of uncertainty attached to this estimate, however, is itself large, so that the use of registry data beyond what is currently available in SEER is important to pursue and will allow more detailed comparisons of age-specific rates of thyroid cancer. Some registries, such as the one in Connecticut, predate the SEER data years by a considerable margin.

This discussion has emphasized the uncertainties inherent in using the NCI (NCI 1997a) report to determine the likely number of excess thyroid cancer cases caused by exposure of the American public to I-131 fallout from the Nevada Test Site. Nevertheless, based on the other data reviewed in this chapter, the committee finds it reasonable to conclude that some excess cases of thyroid cancer have occurred and will continue to occur as a result of the Nevada weapons tests.

TABLE 3.6 SEER Thyroid Cancer Incidence Data. Comparisons of Thyroid Cancer Rates at Age 35-39 Between Birth Cohorts Born in 1938-1942 (Relatively Unexposed) and 1948-1952 (Exposed)

Birth Cohort	Number of 35-39 Year Olds	Number of Thyroid Cancers	Approximate Rate per 100,000 Person-Years	Percentage Excess in Birth Cohort	p-Value for Excess
Far western registries					
1938-1942 (Monitored in 1977)					
Males	209,307	52	4.97	—	
Females	203,509	95	9.34	—	
1948-1952 (Monitored in 1988)					
Males	350,438	59	3.37	−32.2	0.05
Females	344,993	198	11.48	22.9	0.09
Eastern registries					
1938-1942 (Monitored in 1977)					
Males	388,265	59	3.04	—	
Females	407,757	182	8.97	—	
1948-1952 (Monitored in 1988)					
Males	584,760	102	3.49	14.8	0.4
Females	605,706	294	9.71	8.2	0.4

Estimate of the Number of Cases of Thyroid Cancer That Have Already Been Manifested

The population at excess risk of thyroid cancer from I-131 due to fallout from weapons testing at the Nevada Test Site consists of people born between approximately 1940 and 1957. These individuals are now in their early 40's to late 50's in age, and are more than 40 years past their initial exposures. Reported follow-up of the groups considered by Ron and colleagues (1995) is as yet insufficient to be certain that childhood exposure to thyroid radiation results in continued risk for the remainder of one's lifetime (that is, beyond 40 years from time of exposure). Assuming a constancy of excess relative risk obtains, then the distribution of excess thyroid cases would mirror the age specific rates of this cancer. Table 3.7 gives the fraction of thyroid cancer risk manifested by ages 40-60 for males, females, and the sexes combined, with the computations based upon the incidence rates published by SEER.

It should be noted that when the sexes are combined the expected values are closer to those for females than for males because females have a higher background risk of thyroid cancer. Since the birth cohort most at risk is presently between the ages of 45 and 50 years, a reasonable estimate of the fraction of excess cases that have already occurred is about 45 percent. This calculation, as

TABLE 3.7 Fraction of Expected Thyroid Cancer Cases Occurring by a Given Age.

Age (years)	Females (%)	Males (%)	Sexes combined (%)
40	33	20	30
45	42	27	38
50	52	35	48
55	61	45	57
60	69	54	65

are all of the risk calculations in this report, is strongly dependent upon the assumption of a constant excess relative risk throughout a lifetime after early childhood exposure to thyroid radiation. If the excess relative risk declines with attained age, the fraction of cases expressed to date would obviously be different, but in the absence of evidence of such a decline, the committee believes the number, 45 percent, is reasonable as said. Clearly, this value has important implications with regard to the steps that might be invoked to diminish the as yet unmanifested public health consequences of the exposures to I-131.

CONCLUSIONS

Exposure to I-131 as a by-product of nuclear reactions can cause thyroid cancer as shown conclusively by the 1986 nuclear accident in Chernobyl, which resulted in high level exposure for many people. The NCI dose reconstruction model indicates that the level of exposure to I-131 was sufficient to cause and continue to cause excess cases of thyroid cancer. Because of uncertainty about the doses and the estimates of cancer risk, the number of excess cases of thyroid cancer is impossible to predict except within a wide range.

Epidemiological analyses of past thyroid cancer incidence and mortality rates provide little evidence of widespread increases in thyroid cancer risk related to the pattern of exposure to I-131 described in the NCI report. They suggest that any increase in the number of thyroid cancer cases is likely to be in the lower part of the ranges estimated by NCI. The epidemiological analyses are, however, subject to considerable limitations and uncertainties. Given the uncertainties in both the dose reconstruction model and the epidemiological analyses, further epidemiological analyses will be necessary to clarify the extent to which the Nevada tests increased the incidence of thyroid cancer. Pending these studies, it is prudent for DHHS to plan its responses as if excess cases of thyroid cancer have occurred.

Individual-specific estimates of the probability of developing thyroid cancer from exposure to fallout from the Nevada testing program are uncertain to a greater degree than the dose estimates because of the additional uncertainty, in particular, about the cancer-causing effect of low doses of I-131. Nonetheless, the committee concluded that the Nevada weapons-testing program resulted in I-131 exposure that has increased the usual risk of thyroid cancer for some members of the population, mainly those who were very young and drank milk from a back-yard cow or, in particular, a backyard goat. The program of public information discussed in Chapter 5 will inform people about their possible exposure and the risk of thyroid cancer.

ADDENDUM 3: UNDERSTANDING RADIATION RISK FACTORS AND INDIVIDUAL RISK

Radiation cancer risk factors can be expressed in a variety of forms including *Relative Risk* (RR) and *Excess Relative Risk* (ERR). Values of these factors have been discussed in the text and appear in Table 3.4 but both are complex mathematical concepts. These factors can be used to derive a more useful figure for an individual exposed to radiation: that being the *chance* that the person will develop cancer in his or her remaining lifetime following a radiation exposure. This section is intended to elucidate the interpretation and use of these factors for that purpose.

Before proceeding with detailed explanations, it must be noted that there is considerable uncertainty associated with the cancer risk factors as derived from epidemiological studies. Thus, the risk[1] of contracting cancer that one might determine from a calculation should be understood to be only an estimate and the true value of their risk may be several times higher or lower.[2] In addition, the biological response to radiation exposure is different for every individual. Thus, some individuals will be more susceptible to radiation damage and its effects and others less so. Factors influencing individual sensitivity are presently not sufficiently well understood to be accounted for in individual risk calculations.

One useful expression of risk is the *Percentage Lifetime Risk*. *Lifetime Risk* expresses the chance of contracting a cancer within a lifetime (85+ years for a

---

[1]In this discussion, the words *risk, chance, probability*, and *likelihood* can be assumed to have the same meaning.

[2]For the mathematically inclined reader: The extent of this uncertainty will be a combination of the uncertainty in the dose estimate and the uncertainty in the cancer risk factor derived from epidemiological studies. The uncertainty in the dose estimate will vary with location, with higher uncertainties in those western states nearer to the Nevada Test Site (a factor of 6 to 20 either side of the estimated geometric mean dose) and lower uncertainties in the eastern United States (a factor of 3 to 10 either side of the estimated geometric mean dose). The uncertainty in the cancer risk factor has been estimated to be approximately a factor of 3.6 on either side of the central estimate (the central estimate of risk is an Excess Relative Risk of 7.7 per gray [Gy]). The overall uncertainty in the *Percentage Lifetime Risk* estimate for an individual is large and could be 5 to 30 times depending on location of residence at the time of exposure. This means that the true *Percentage Lifetime Risk* could be much smaller (by 5 to 30 times) or much larger (by 5 to 30 times).

newborn) or within the remainder of life following a radiation exposure. Because all types of cancers can occur in the absence of radiation, there is a background incidence rate even when no exposure has taken place. Thus, the *Percentage Lifetime Risk* is never zero, even in the absence of radiation exposure above that from natural background radiation.[3] In the case of thyroid cancer, about 0.25 percent of males (1 in 400) and 0.65 percent of females (1 in 153) will have the disease within his or her lifetime for reasons unrelated to radiation exposure (above that from natural background radiation). Thus, without any radiation exposure above that from natural background radiation, the *Percentage Lifetime Risk* for males and females is 0.25 and 0.65, respectively (see line 6 of Table 3.4).

The *chance* of a single person developing a disease is a difficult concept for most people to understand because one either contracts the disease or does not. *Lifetime Risk* can perhaps be better explained by defining it to express the number of people out of each 100 similar people that would develop the disease. The same numerical value of *Percentage Lifetime Risk* can apply equally well to the chance for an individual to develop cancer, or to the proportion of people out of each 100 that will develop cancer. The interpretation used is largely a matter of individual preference.

In the case of thyroid cancer caused by radiation, adults of 20 years of age and greater at time of exposure are considered to be at negligible risk. Thus, their *Lifetime Risk* (see column 6 of Table 3.4) is equal to the background incidence of this disease in an unexposed American population (0.25 percent males, 0.65 percent females). The *Percentage Lifetime Risk* from the Nevada Tests for other age cohorts is shown in Table 3.4 based on the average radiation dose noted in column 3 as computed by NCI. The *Percentage Lifetime Risk* is, in general, higher for larger radiation doses, higher when exposure occurs at younger ages, and higher for females compared to males at the same age and the same radiation dose.

In Table 3.4, adults over 20 years of age (last line of the table) can serve as a reference group to which other age cohorts are compared. Using that simple idea, a *Relative Risk* figure (see column 6 of Table 3.4) can be calculated, which simply expresses a multiple of the risk to an unexposed person or the adult. In Table 3.4 for example, the *Relative Risk* for individuals of age 15-19 years at time of exposure is 1.01 or 1 percent higher than the adult groups for the same average dose as the adults received. This small increase in the *Relative Risk* does not significantly affect *Lifetime Risk* which is about 0.25 percent, the same as for older peers.

As mentioned earlier, the *Relative Risk* increases with increasing dose but is also greater for younger ages of exposure. That effect can also be seen in Table 3.4, which gives higher *Relative Risks* (always in relation to the adult cohort) for the younger groups who also had higher average doses from the Nevada tests. Higher *Relative Risks* result in proportionally higher *Percentage Lifetime Risks*.

---

[3]The thyroid gland would receive on average about 0.001 Sv annually from natural ionizing radiation in the environment.

For example, those less than 1 year of age at exposure (who received on average 10.3 rad) have a *Relative Risk* of 1.67. This is the same as saying the risk is 1.67 times the risk of adults (who received an average dose of 1.8 rad). Consequently the *Percentage Lifetime Risk* for the less than 1 year old group is 1.67 times that of adults, equal to 0.42 percent (males) or 1.07 percent (females).

In lieu of presenting further mathematical detail necessary to manipulate the risk factors, it appears more useful to further discuss only the *Percentage Lifetime Risk*. This value may be obtained directly from Table 3.4 if certain assumptions can be made. Specifically, the table presents the *Percentage Lifetime Risks* for individuals who received a radiation dose equal to the average for his or her age cohort as estimated by NCI. Column 5 of Table 3.4 gives those risk values given the assumptions used. The highest risk for a male is 0.42 percent, which is for an exposure of about 10 rad at one year of age. The highest risk for a female would be 1.1 percent.[4] The risk of 0.42 percent (male) or 1.1 percent (females) equates to about 1 in 240 males of his age cohort or 1 in 92 females developing thyroid cancer sometime in her life if lived to age 85+.

The disease rates computed above for a 10 rad exposure to a 1 year old are 67 percent higher than the background incidence, which is 1 in 400 males or 1 in 173 females developing thyroid cancer sometime in his or her life. By way of comparison, women have about a 1 in 8 *Lifetime Risk* to age 85 of being diagnosed with breast cancer and men about a 1 in 5 *Lifetime Risk* of being diagnosed with prostate cancer.

For the *Lifetime Risks* presented in Table 3.4 to be applicable to any individual, that person must have received the national average dose for his or her age cohort. For a person to determine the dose with any confidence would require extensive calculations, even then the uncertainty would be great. For those individuals highly concerned about their individual risks, determining relevant doses specifically for themselves is a difficult undertaking. At present, there are few alternatives to completing complex calculations though an alternative method is suggested in Addendum 5.

One method for a person to estimate his or her personal risk would begin with obtaining an estimate of the individual *total* dose (from all tests) from the NCI Web site. That process is difficult and the concerns expressed in this report about the uncertainty of calculated doses should be borne in mind. Under the assumption that the computed dose accurately represents the exposure of the individual, one's *Percentage Lifetime Risk* could then be estimated by the following steps:

1) Obtain the Percentage Lifetime Risk from Table 3.4 for the age cohort.

---

[4]For females, the *Percentage Lifetime Risk* is estimated to be 2.6 times greater than males at any dose.

2) Subtract 0.25 (males) or 0.65 (females) (the *Lifetime Risk* for the adult group) from the value found in step 1.

3) Calculate the ratio of the one's dose estimate to the average dose for the age cohort.

4) Multiply this ratio by the value from step 2 above.

5) Add the adult value for *Percentage Lifetime Risk* (0.25 for males or 0.65 for females) to the number from step 4 above.

The number calculated by the above five steps would be an estimate of one's individual *Percentage Lifetime Risk*. It could be interpreted as the percent chance of developing thyroid cancer during their lifetime to 85+ years of age, or the proportion of 100 people who would develop the disease. For example, if the person was less than 1 year old at time of exposure and his or her individual dose estimate was about 50 rad, he or she would calculate a *Percentage Lifetime Risk* of about 1.1 (males) or 2.9 (females), meaning that the chance of developing thyroid cancer by age 85 would be about 1 percent (males) or 3 percent (females). Equivalently, for this 50 rad dose, it would mean that 1 of every 100 males or 1 of every 35 females like themselves would contract the disease because of the radiation exposure. Relatively few individuals likely received doses as high as 50 rad, thus relatively few persons would have chances of developing thyroid cancer this great.[5]

Two final points are worthy of note in this discussion of risk estimates. First, for those who are diagnosed with radiation related thyroid cancer, the prognosis, as previously discussed, is good. Second, the individuals that are diagnosed with thyroid cancer can have a relatively high probability that the disease was caused by radiation exposure even if the likelihood that they would get the disease was low. For example, the probability that a cancer is a result of radiation exposure (this probability is termed "probability of causation" or PC) (NIH 1985) can be calculated as [Relative Risk − 1]/[Relative Risk].[6] Thus, for the child having received an actual dose of 10.3 rad (see Table 3.4), the PC would equal (1.67 − 1)/1.67 = 0.40. This corresponds to a 40 percent chance the cancer was a result of radiation exposure. This can be compared with a chance of only 0.42 percent (male) or 1.07 percent (female) that they would have developed the disease during their lives following a 10.3 rad exposure at 1 year of age. The PC becomes increasingly higher for larger radiation doses.

---

[5]The primary exception would be individuals who routinely drank goats' milk. NCI estimated the number of such individuals in the United States who were in childhood at time of exposure to have been about 20,000. These persons could have received doses greater than 100 rad at many locations.

[6]There are alternate ways to write this equation, all are equivalent: PC = (RR − 1)/RR = ERR/RR = ERR/(1 + ERR).

4

Implications for Clinical Practice and Public Health Policy

Efforts to measure and understand the implications of the exposure of Americans to radioactive fallout from the atomic bomb tests at the Nevada Test Site in the 1950s raise a variety of clinical and public health issues including directions for further research, strategies for informing the public about risk, and policies for disease screening. This chapter focuses on thyroid cancer screening for people thought to be at higher-than-average risk for thyroid cancer from exposure to iodine-131 (I-131) fallout. It also briefly considers screening for other thyroid disorders. The final sections consider clinical and public health policies based on a review of the scientific literature on the accuracy, benefits, and harms of screening.

In public health terms, screening for thyroid cancer is a form of secondary prevention.[1] Like screening for other cancers, the goal is to detect disease in people without symptoms so that they can be treated early to reduce mortality and morbidity. Routine screening is directed at the population generally; targeted screening seeks people who have a higher-than-average risk of developing a disease. In either case, a screening program has value only when earlier detection of a disease results in earlier treatment that improves outcomes for the screened population.

The appeal of screening as a means of reducing the burden of disease is powerful, and much has been claimed for a large array of tests that screen for

---

[1]Primary prevention aims to eliminate or reduce health threats (e.g., by treating waste water) and to make people less susceptible to such threats (e.g., through vaccination). Tertiary prevention involves treatment and management of existing illness (e.g., through drug therapy for diagnosed heart disease) to reduce the threat of complications or progression.

diseases causing premature morbidity and mortality. The U.S. Preventive Services Task Force has, for example, examined and made recommendations about screening for 53 conditions, including heart disease, several kinds of cancer, infectious diseases, prenatal disorders, and sensory problems. As these and other recommendations and analyses make clear, clinical epidemiologic research confirms the value of some screening tests but does not support claims for others (USPSTF 1996; Russell 1994; Eddy 1991).

Efforts to evaluate screening strategies and to develop evidence-based recommendations for screening can generate considerable controversy. Science-based conclusions (especially when the conclusion is that the evidence for screening is negative, inconclusive, or lacking) can conflict with the understandable public desire to believe that a particular screening test will save lives. A case in point is the controversy over a recommendation from an NCI consensus panel that screening for breast cancer in women ages 40 to 49 should be a matter for women to decide with their clinicians rather than a routinely advised practice (Begley 1997; Eddy 1997; Ransohoff and Harris 1997). Following protests and criticism from some advocacy groups and some members of Congress, a different NCI panel (the National Cancer Advisory Board) recommended routine breast cancer screening for this age group (Taubes 1997) even though many scientists think that evidence is still inadequate to support general screening in this age group.

The discussion in this chapter builds on sections in other chapters of this report that have examined who is potentially at risk of thyroid cancer from I-131 exposure, how great the risk is, and how communication with the public should be structured. It also draws on the literature review and analyses presented in the background paper commissioned for this study (Appendix F). The discussion here also builds generally on the principles of evidence-based clinical practice and public health policy. It recapitulates some of the information on thyroid cancer presented earlier, so that this chapter can be read independently. This chapter reviews

- The concepts and principles for screening recommendations.
- The burden of illness associated with thyroid cancer.
- The benefits and harms of screening.
- The tests used for screening.
- The evidence about test accuracy and the benefits of early detection.
- The screening recommendations of other groups.
- This study's conclusions and recommendations.

PRINCIPLES FOR SCREENING RECOMMENDATIONS

The specific conclusions about thyroid cancer screening for exposed persons were developed by the Institute of Medicine (IOM) committee that was described in Chapter 1. In developing its recommendations, the committee examined the scientific literature (including the literature review by Eden, Helfand, and Mahon in Appendix F), considered position statements and guidelines developed by other groups and individuals, and benefited from the discussion of diverse experts at a March 17-18, 1998, workshop convened by the committee (see Appendix A for the agenda and participants). In addition, the committee reviewed information from the public meetings, analyses, and deliberations of the National Research Council committee that considered the I-131 thyroid doses from the Nevada tests, the cancer risk posed by I-131 exposure, and the approaches to communicating with the public about exposure and risk.

Criteria for Clinical Recommendations

In developing guidelines for thyroid cancer screening for people potentially exposed to I-131 fallout from the Nevada tests, the IOM committee began with several broad principles. Consistent with the established and recognized mission of the IOM and the National Research Council, the guidelines would be based on careful review and assessment of the scientific evidence. They would be intended to assist practitioner and patient decisions about appropriate health care for specific clinical circumstances and to inform policy decisions about public health strategies.

Recommendations set forth by the IOM and others (CDC 1996; USPSTF 1996; McCormick and others 1994; IOM 1992; Eddy 1991) have proposed a number of additional criteria for guidelines for clinical practice.

First, the *disease* targeted for screening should be

- Important in terms of prevalence, incidence, and mortality or morbidity (disease burden).
- Amenable to treatment that produces better outcomes (benefits balanced against harms) than observation alone or no treatment at all.
- More successfully treatable (e.g., condition cured, progression slowed) if detected at an earlier stage than would be possible without screening (e.g., before symptoms are evident).

Second, screening recommendations should be based on evidence related to the technical characteristics of a test used for screening. A *screening test* should be

- Reliable, accurate, and safe in detecting the disease earlier than would be possible under the conditions of usual care, as indicated by the test's sensitivity, specificity, and positive predictive value (see Addendum 4A for definitions and an illustration of the effect of disease prevalence on screening test performance).
- Projected to have benefits (e.g., improved life expectancy or quality of life) that outweigh harms (e.g., possible unnecessary diagnostic work-ups or surgery for false-positive results or nonprogressive cancers).

Third, for screening to be implemented successfully, a *screening strategy* should be

- Acceptable to the public (e.g., involve reasonable burdens or harms as perceived by those to be screened).
- Acceptable to clinicians (e.g., not unreasonably burdensome to undertake).
- Effective and feasible to implement in normal practice.
- Cost-effective (e.g., have a cost per life saved or quality-adjusted life-year gained that is comparable to or less than that provided by other accepted interventions).

The last of these common criteria for screening—cost-effectiveness—was *not* considered by the committee in its analyses nor was such analysis specified in the charge to the IOM from the National Cancer Institute. Clinical evidence and considerations alone determined the committee's conclusions. The committee, however, noted that if evidence does not support claims that an intervention is clinically effective or that its benefits outweigh its harms, then the further step of cost-effectiveness analysis makes no sense.

The committee's criteria for making recommendations about thyroid cancer screening for those exposed to I-131 from the Nevada tests can be depicted as an evidence pyramid (Figure 4.1). The lowest tier involves evidence of a population health problem; next are the availability of effective treatment for the disease and of accurate and feasible screening tests; a yet higher tier involves evidence that early detection through screening improves outcomes; and at the top of the pyramid is evidence that benefits exceed harms. Recommendations for screening apparently health populations are generally held to a higher standard of effectiveness than recommendations for treatment in people with evident disease or injury.

To make a positive recommendation for screening for thyroid cancer in people exposed to I-131 from the Nevada nuclear tests, the committee determined in advance 1) that a chain of evidence was required that exposure to I-131 from fallout increases the risk of thyroid cancer; 2) that effective treatment is

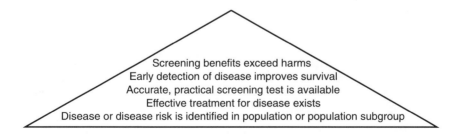

FIGURE 4.1 Pyramid of evidence for national screening policy.

available for the disease; 3) that screening tests for thyroid cancer are reliable, accurate, and practical in detecting disease earlier than would occur with usual care; 4) that early detection of thyroid cancer by screening improves treatment outcomes; and 5) that the potential benefits of screening outweigh its potential harms. The strength of either a positive or a negative recommendation would depend on the strength of the evidence on these four points. For example, if evidence suggested that screening was efficacious but that good results depended on the skill or experience of the clinicians performing the screening tests, then a recommendation might describe when referral to a more experienced clinician should be considered. A committee recommendation about an intensive public health program to encourage or pay for screening would have to note (as described above) that the committee did not analyze the cost-effectiveness of thyroid cancer screening compared with other screening strategies of demonstrated value.

BURDEN OF ILLNESS

General Burden of Mortality and Morbidity

Chapter 3 presented data for thyroid cancer in the United States. To recapitulate in the context of this discussion of screening policy, thyroid cancer is uncommon and rarely life-threatening. The overall age-adjusted incidence for all forms of thyroid cancer is 4.9 per 100,000 population (unless otherwise indicated, data are from NCI 1997b), and it accounts for about 1 percent of cancers diagnosed each year and about 0.2 percent of cancer deaths (an estimated 1,200 out of 564,800 in 1998) (ACS 1998a). The majority of these cancers are papillary carcinomas, which is also the form of thyroid cancer linked to I-131 exposure.

Naturally occurring thyroid cancer is considerably more common in women than it is in men, 6.9 cases per 100,000 for women versus 2.8 for men (NCI 1997b), and it is more frequent in whites than in blacks.[2] For men, the incidence

---

[2]These data come from SEER (NCI 1997b). SEER stands for the Surveillance, Epidemiology, and End Results Program, which is part of the National Cancer Institute. The SEER program was established following 1971 legislation directing NCI to collect, analyze and disseminate data useful in preventing, diagnosing, and treating cancer. The SEER data reported here were collected from 9 or 11 designated cancer registries. For all years, the registry reporting areas include Connecticut, Iowa, New Mexico, Utah, Hawaii, Detroit, San Francisco-Oakland, Atlanta, and Seattle-Puget Sound with data from Los Angeles and San Jose-Monterey included for the most recent period. These registries cover an estimated 9.5 (9 areas) or 13.9 percent (11 areas) of the U.S. population. The registries abstract records for all resident cancer patients seen in hospitals inside and outside the geographic area; search other records (e.g., private laboratories) to identify additional patients; abstract death certificates for residents with a cancer diagnosis listed regardless of site of death; and follow most identified living patients who have a cancer diagnosis. Information abstracted includes patient demographics and primary disease site, morphology, diagnostic confirmation, extent, and initial therapy.

of thyroid cancer increases fairly steadily until age 70; for women, incidence peaks in middle age. The median age at diagnosis for all forms of thyroid cancer is 43 years for women and 50 years for men. For those who die of all forms of thyroid cancer (a small proportion of those diagnosed), the median age at death is 75 years for women and 69 years for men.

Survival rates for this disease are high. The 10-year cancer-specific survival rate for persons with papillary carcinoma, the form linked to radiation exposure, is estimated at 95 percent and the 30-year survival rate is estimated at 90 percent (Wang and Crapo 1997; Mazzaferri and Jhiang 1994). Women's survival rates are somewhat higher than rates for men (96 percent for women versus 92 percent for men at 5 years for all thyroid cancers combined). Survival rates are somewhat lower for blacks than for whites (around 88 percent at 5 years for all thyroid cancers combined).

Age-adjusted mortality rates for all forms of thyroid cancer are 0.4 per 100,000 for women and 0.3 per 100,000 for men. (By way of comparison, the mortality rate for lung cancer is 32.8 and 73.2 per 100,000 for women and men respectively; for breast cancer in women, the mortality rate is 26.4 per 100,000 and for men, the mortality rate for prostate cancer is 26.9 per 100,000.) Although relatively low at all ages, mortality rates rise after about age 45, an age already reached or being approached by the cohort potentially at risk of thyroid cancer from exposure to radioactive fallout from the Nevada tests.

Chapter 3 noted that thyroid cancer is among the cancers that have increased in incidence but decreased in mortality rate over the past 30 years. The contrast between the incidence and mortality trends has been attributed to more sophisticated detection technologies (ultrasound for nodules and FNA biopsy for cancer) and more complete diagnostic reporting (Wang and Crapo 1997). The disease accounts for about 1 percent of all cancers and about 0.2 percent of cancer deaths (an estimated 1,200 of 564,800 in 1998, according to the American Cancer Society 1998a).

Mortality does not represent the only burden of thyroid cancer. Symptoms of the disease itself, the processes of diagnosis and treatment, and the consequences of and follow up after treatment also must be considered. Symptoms for treatable disease are relatively uncommon but can include hoarseness and difficulty swallowing or breathing. Diagnosis by fine-needle aspiration (FNA) biopsy can cause anxiety and minor physical discomfort, as described later in this chapter and can lead to overtreatment for small lesions.

Surgical treatment for clinically significant papillary thyroid cancer is generally considered to provide effective treatment and improve long-term survival (Mazzaferri and Jhiang 1994; Mazzaferri 1993b; Hay 1990). Some controversy, however, remains about specific surgical strategy.[3] Surgical treatment involves

---

[3]See, for example, Cady and Rossi (1988), Shaha and others (1995) on the case for subtotal thyroidectomy and Mazzaferri (1996), Schlumberger (1998), and Demure and Clark (1990) on the case for near-total thyroidectomy.

inpatient surgery that typically removes nearly all of the thyroid gland (leaving a small part of the gland near the entrance of the laryngeal nerve into the larynx) (Falk 1997). Surgical scars are generally visible on the neck above shirt or blouse collars, which may bother some people and not others. If a thyroid cancer, other than a very small one is found, surgery is often followed by radiation treatment with I-131 to destroy remnant thyroid tissue. After total or near-total thyroidectomy, patients require lifetime thyroid hormone replacement therapy, which if properly prescribed and monitored, does not present significant risk of adverse side effects.

In addition to a very small risk of complications from anesthesia, identified surgical risks include permanent damage to the laryngeal nerve (with resulting severe hoarseness) and inadvertent removal of the parathyroid glands (with medication required to treat subsequent problems related to hypocalcemia, which can include muscle tremors, cramps, and seizures). A recent review of seven studies, involving a total of 1,754 patients, reported permanent hypoparathyroidism rates of 0.8 to 5.4 percent (weighted mean, 2.6 percent) for those undergoing total thyroidectomy; for subtotal thyroidectomy, the weighted mean was 0.2 percent (Udelsman 1996). The review also reported permanent recurrent laryngeal nerve injury in 0 to 5.5 percent (weighted mean 3.0 percent) for total thyroidectomy patients and in 1.1 to 3.2 percent (weighted mean, 1.9 percent) of those undergoing the less extensive procedure. Although the committee is not aware of specific research on the relationship between thyroidectomy complication rates and surgical skill or experience, the technical difficulty of thyroid surgery, especially for total or near-total thyroidectomy, is such that lower complication rates might reasonably be expected from surgeons and centers that perform higher volumes of procedures.[4] Patients who undergo surgery for large invasive tumors or who have extensive lateral lymph node dissections (i.e., standard radical neck dissection) combined with total thyroidectomy are at higher risk for complications, as are older patients who are likely to have comorbid conditions (Mazzaferri and others 1977).

Exposure to I-131 and Risk of Thyroid Problems

The NCI report links higher exposure to I-131 to a combination of geographic location, age at exposure, and milk consumption, especially milk from "backyard" cows and, particularly, goats. As discussed in the NCI report and in Chapter 2 of this IOM/NRC report, the NCI's estimates of the American public's exposure to I-131 from the 1951-1962 nuclear tests in Nevada are characterized by

---

[4]Chen and others (1996) discuss outcomes for parathyroidectomy specifically. General discussions of the procedure appear in Hughes and others (1987), Banta and others (1992), Grumbach and others (1995), Imperato and others (1996), Tu and Naylor (1996), and Wennberg and others (1998).

substantial uncertainty. This uncertainty, which is inevitable given data limitations related to geographic measurements of fallout and reconstruction of milk consumption information, applies to general characterizations of total population exposure but it creates the greatest problems for making individual estimates. Accurate estimates of a specific individual's exposure are usually not possible.

Chapter 3 concluded that evidence supports a link between exposure to radioactive iodine from fallout and papillary thyroid cancer. The major uncertainty of this relationship would be at the lower exposure levels that characterized most Americans' exposure from the Nevada weapons testing program. For the average American woman, the probability of being diagnosed with thyroid cancer by the age of 95 is roughly 1 chance in 150 without exposure and 1 in 90 with exposure to 0.1 Gy (10 rad). For the average man, the corresponding probabilities are 1 in 400 without exposure and 1 in 150 with exposure. As noted earlier, papillary thyroid cancer, the form related to radiation exposure, is rarely life threatening.

Nonmalignant Thyroid Disease Associated with I-131 Exposure

In addition to thyroid cancer, Chapter 3 investigates whether evidence links I-131 exposure to other thyroid disorders, which include thyroid nodules, hypothyroidism, hyperthyroidism, and autoimmune thyroiditis and goiter. The review in Chapter 3 cites considerable evidence of links at moderate and high exposures but little evidence suggesting a link between hypothyroidism or hyperthyroidism and I-131 doses in the range experienced by most of those exposed to fallout from the Nevada nuclear tests. For example, a study of Utah "downwinders" exposed to atomic weapons test fallout after birth (mean dose of 0.098 Gy [9.8 rad] with a maximum of 4.6 Gy [460 rad]) showed no excess of nonmalignant thyroid disease (Kerber and others 1993).

Given the conclusions in Chapter 3, the IOM committee did not investigate the disease burden of nonmalignant disease or the evidence base for screening for such disease in people exposed to I-131 from the Nevada nuclear tests. The results of the Hanford Thyroid Disease Study (see www.fhcrc.org/science/phs/htds) may clarify the extent to which childhood exposure to low doses of I-131 is linked to thyroid problems in adulthood. This clarification should, in turn, clarify whether it is warranted for DHHS to examine the evidence on screening for nonmalignant disease.[5]

THYROID CANCER SCREENING AND DIAGNOSTIC OPTIONS

Screening for thyroid cancer may involve two steps. For the first step, screening for thyroid nodules, the options reviewed by the committee are physical pal-

---

[5] As this report was nearing public release and after the committee had concluded its deliberations, the American College of Physicians published new recommendations for screening for benign thyroid disease (Helfand and others 1998; Helfand and Redfern 1998).

pation and ultrasound examinations of the neck (Charboneau and others 1998). If these tests detect a large nodule (1.5 cm or larger), the second step, FNA (fine needle aspirate) biopsy of the nodule, is usually performed. Observation (with periodic examinations) will be the approach for most of those who have small (less than <1.5 cm), otherwise unsuspicious lesions detected by palpation or ultrasound. Most such lesions, often discovered during ultrasound examinations for other conditions (e.g., carotid artery disease, parathyroid problems), are benign and even those that have cancer cells will usually not become life- or health-threatening. Some patients are referred directly for surgery following detection of a nodule, although this is not standard or recommended practice (AACE 1997a; Tan and Gharib 1997; Ezzat and others 1994; Mazzaferri 1993b).

Physical palpation of the thyroid[6] is performed in conjunction with a visual examination of the neck and a patient history that documents possible risk factors such as age, sex, family history of thyroid disease (especially papillary or medullary cancer), personal history of previous head or neck irradiation or thyroid problems, or symptoms such as hoarseness or difficulty swallowing (Gharib 1997; Belfiore and others 1995; Mazzaferri 1993b). Palpation seeks to detect nodules in the thyroid, and if they are found, to assess their size, number, firmness, and adherence to adjacent structures. The size and firmness of nearby lymph nodes are also checked by palpation. For asymptomatic, average-risk people, routine screening for thyroid cancer through palpation or ultrasound examination of the thyroid gland is not recommended (USPSTF 1996).

Ultrasound examination of the thyroid involves noninvasive scanning of the thyroid gland with a machine that creates images by directing high-frequency sound—ultrasonic waves—to the gland and using computer algorithms to translate information about sound transmission or reflection (echoes) into two-dimensional images (Tan and others 1995; Ezzat and others 1994; Hsiao and Chang 1994). Ultrasound can detect many small nodules (less than 1 to 1.5 cm) that are not usually detectable by palpation and that will usually not progress to cause problems. Little is known about the natural history of these small nodules and the occasional cancers, including what causes a few of them to progress. Rather than further tests or therapy, such nodules generally warrant observation (Gharib and Mazzaferri 1998). Ultrasound examination also can provide additional information to further characterize nodules, including whether calcification is present and whether the nodule is solid or cystic (Watters and Ahuja 1992). Although such information—in combination with information from physical examination—can raise or lower the probability that a nodule is cancerous, ultrasound information is not by itself diagnostic.

---

[6]This is a more focused examination of the thyroid than the physical examination of the neck that physicians may perform as part of a routine physical. The latter examination also includes palpating cervical nodes, listening for carotid bruits with the stethoscope, checking range of motion in the neck, feeling vertebral processes, and other elements.

To determine whether a detected thyroid nodule is benign or malignant, the current procedure of choice is FNA biopsy (Garcia-Mayor and others 1997; Gharib 1997; Lin and Huang 1997; Ashcraft and Van Herle 1981). Typically, a fine gauge needle is inserted into the nodule (or nodules) in several places and a sample drawn into the needle and an attached syringe. If the nodule is not readily palpated, the biopsy can be guided by ultrasound, although this technology is not available everywhere and can lead to the identification of additional, small nodules, few of which are likely to cause problems (Lin and Huang 1997; Taki and others 1997). Minor localized pain is common during and after the biopsy, and, rarely, a clinically apparent hematoma can occur at the FNA site (Gharib and Goellner 1993; Ashcraft and Van Herle 1981).

Following the biopsy, the aspirated material (tissue and blood) is placed on slides or filtered through mesh and applied to slides, stained with various dyes, and then microscopically examined. Pathologists evaluate the sample on the slides and classify the lesion. In addition to positive or negative results, FNA samples can be categorized as indeterminate or unsatisfactory. For results categorized as indeterminate, the pathologic features of the cells are ambiguous and cannot be readily categorized as benign or malignant. Patients with indeterminate biopsies are usually referred to surgery. Of satisfactory samples, about 10 percent are likely to be classified as indeterminate.

Unsatisfactory samples provide insufficient or inadequate material for evaluation and often prompt repeated biopsies aimed at securing an adequate sample for further analysis (LiVolsi 1997). Unsatisfactory FNA samples are obtained 10-15 percent of the time, even in the best-performing centers. The rate of unsatisfactory samples depends on the training and skill of the operator (e.g., endocrinologist, head-and-neck surgeon, radiologist, family practitioner), the number of sites sampled, preservation practices, the skill of the pathologist or cytopathologist, and the size of the nodule being biopsied (Merchant and Thomas 1995; Haas and Trujilla 1993; Piromalli and Martelli 1992). A screening approach that led to more FNA biopsies of smaller nodules (<1-1.5 cm) would likely generate more inadequate samples and, in turn, more repeat biopsies. Although no precise figures are available and recommendations vary, some patients, perhaps even a majority, with inadequate FNA samples will be referred for surgery. Addendum 4B considers the implications of indeterminate or unsatisfactory FNA biopsy results in more depth.

ACCURACY OF SCREENING AND FOLLOW-UP TESTS

Palpation and ultrasound examinations are widely accepted as safe, low-risk procedures (Gharib 1997; Mazzaferri 1993a; Ashcraft and Van Herle 1981). Concern about their use is not related to the direct risk of the procedures themselves but to their relative inaccuracy, particularly the probability of their producing false-positive results or identifying many very small nodules and cancers that are not likely to cause harm. Because the likelihood of finding thyroid cancer (pretest prevalence) is low, even in most populations exposed to iodine-131, the great majority of positive test results for nodules will be false-positive test results for cancer when the nodule is biopsied. A detailed review of evidence about the accuracy of palpation and ultrasound in detecting thyroid nodules and of FNA in detecting thyroid cancer is presented in the "Screening for Thyroid Cancer" background paper (Appendix F).

Palpation

The standard for reporting sensitivity in the detection of thyroid nodules is ultrasound. Because ultrasound cannot definitively discriminate benign from malignant nodules, the test results in nodule—not cancer—detection. Research has not compared the accuracy of different approaches to palpation (e.g., whether it is best done from the patient's front or back) (Tan and Gharib 1997), but more experienced examiners are likely to produce more accurate results (Jarlov and others 1993; Jarlov and others 1991). Accuracy can be affected by the size, consistency, or location of nodules (e.g., on the front or back surface of the thyroid) and by a patient's physical characteristics (e.g., length and thickness of the neck).

The sensitivity results from three studies of palpation ranged from 0.10 to 0.31 in detecting nodules identifiable with ultrasound (Table 1 in Appendix F). The study that reported 0.31 overall sensitivity found sensitivity of 0.45 for detection of nodules larger than 0.5 cm and 0.84 for nodules larger than 1 cm (Ezzat and others 1994). A study of palpation in a radiation-exposed population reported sensitivity of 0.89 for detection of nodules larger than 2 cm and 0.83 for those between 1 and 2 cm (Mettler and Williamson 1992). Other studies of the sensitivity of palpation for detecting larger nodules gave poorer results. For example, one study reported sensitivity of only 0.42 for nodules larger than 2 cm (Brander and others 1992). The U.S. Preventive Services Task Force (USPSTF 1996) reports specificity levels for nodule detection of 0.93 to 1.0 from three studies that compare palpation with ultrasound.

Ultrasound

Ultrasound examination is the reference standard for nodule detection, so its sensitivity and specificity for this purpose are not reported. It very commonly reveals thyroid nodules in people without symptoms who are being examined for other purposes. For example, four U.S. studies of ultrasound examinations report nodules in 13 percent (Carroll 1982), 41 percent (Horlocker and Jay 1985), 50 percent (Stark and others 1983), and 67 percent (Ezzat and others 1994) of those examined. Current ultrasound technology does not allow reliable and accurate discrimination between cancerous and noncancerous nodules.

When ultrasound results are compared against FNA results for specificity in detecting cancer (rather than nodules), studies report a high rate of false positives, that is, nodules without cancer. In patients examined by ultrasound for hyperparathyroidism, Horlocker and colleagues (Horlocker and Jay 1985) found thyroid nodules in 41 percent, but cancer was detected in only 2 percent of the 689 patients referred for surgical neck exploration. The trend of increasing nodule detection with age found by Horlocker and colleagues is fairly close to pathology findings at autopsy reported in a much earlier study (Mortensen and others 1955).

Fine-Needle Aspiration Biopsy

A recent set of guidelines on cytopathology for FNA specimens from thyroid nodules described FNA of the thyroid as "principally a triage procedure" to differentiate patients who need surgery from those who do not (Papanicolaou 1996). It noted, however, that "in experienced hands [FNA] may also be diagnostic for certain thyroid lesions" such as papillary and other carcinomas (p. 710).

The standard for reporting sensitivity and specificity of FNA is findings at surgery. Although most patients with negative or benign FNA results do not have surgery (and thus no comparison against this standard is possible), several longitudinal studies of patients with negative test results indicate that such results are reliable (Tan and Gharib 1997). For pathologists who are experienced at interpreting FNA biopsies of thyroid nodules, false-positive rates of 3 percent and false-negative rates of 2 percent have been reported (Gharib and Goellner 1993; Hall and others 1989; Boey and others 1986). As shown in Table 2a and 2b in Appendix F, however, sensitivity and specificity rates can vary depending on how analyses treat unsatisfactory or indeterminate FNA results. (See Addendum 4B for further discussion of different interpretations of indeterminate and unsatisfactory samples.) Not only will many patients referred for FNA turn out not to have thyroid cancer, but many will likely be referred for surgical resection of the thyroid because of unsatisfactory or indeterminate results.

BENEFITS AND HARMS OF SCREENING FOR THYROID CANCER

General Considerations

Most discussions of cancer screening emphasize potential benefits, but potential harms also must be considered and weighed. The primary benefit sought from screening is early detection of asymptomatic disease and treatment at a stage that permits improved outcomes including extended life (not just a longer period with a diagnosis), reduced morbidity, and better quality of life. Those whose screening results are negative (no disease detected) may feel comforted, particularly if they view themselves as being at special risk of the disease.[7] Even

---

[7]For breast cancer screening, Ransohoff and Harris (1997) have suggested that this reassurance may be out of proportion to a change in probabilities of cancer before and after a negative screening result. They suggest more reassurance might derive from a better understanding of actual risk.

when early detection of a disease is not matched by effective treatment (e.g., as is now true for some genetic screens), some people might believe themselves better off with an early diagnosis because they could plan more realistically for the future. More generally, a screening program for a problem linked to past government actions may be viewed as a government admission of responsibility for harm, which may provide a psychological and financial benefit (in the form of compensation) for those who believe themselves harmed.

The potential harms of cancer screening as experienced and perceived by patients are not necessarily well defined or understood. For those who have false-positive screening results (i.e., no cancer is actually present), the harms can include anxiety, inconvenience, and physical discomfort attendant to follow-up diagnostic testing; unnecessary surgery or other treatment—and associated morbidity or mortality—either initially or when a follow-up diagnostic test is inaccurate or indeterminate; and self-labeling or external labeling (e.g., by insurers, employers, family) of a person as high risk or vulnerable. Even those with true-positive results can suffer if early detection and treatment do not affect outcomes. If they have a nonprogressing cancer, they may experience needless anxiety and treatment, and even if their cancer progresses, they may experience a longer period of anxiety about their condition or endure more years of treatment and adverse treatment-related side effects without any better outcomes than if they had not been screened. Those who have false-negative screening results can subsequently be less alert to early symptoms of disease and could delay seeking care.

As far as the committee is aware, there are very few studies of patient perceptions of quality of life for health states associated with different outcomes of screening, diagnosis, or treatment for specific cancers. It is possible that experts who assess screening tests either over- or underestimate the benefits and harms as perceived by patients who consider and undergo screening—especially patients who see themselves as being at higher risk of a disease. Some research suggests that the public may have a high tolerance for false-positives (Schwartz and Woloshin 1998). This research also suggests that people may be unaware that screening may identify many cancers that will not progress even without treatment and that people want to be informed and to factor this information into their decisionmaking process.

Evidence of Benefits from Early Detection through Thyroid Cancer Screening

A major difficulty faced by the committee in considering its recommendations on screening for thyroid cancer was the absence of sound clinical research evaluating whether early detection of the disease through screening of asymptomatic people provides benefits in the form of longer life, reduced morbidity, or improved quality of life and whether such benefits outweigh any harms generated by screening. Rigorous prospective, randomized clinical studies of screening benefits are generally complicated, expensive, and time consuming. Ideally, they would randomly assign asymptomatic individuals to be screened or not screened and then track subsequent survival or other outcomes. No such studies of thyroid cancer screening have been published.

Thyroid cancer screening is a difficult subject for rigorous clinical research because the disease is uncommon, progresses slowly, and is rarely fatal. Because any change in survival would likely be quite small and not seen for many years, a clinical trial of screening would have to be both very large and very long. Indeed, by the time a trial was completed, the question of screening probably would be irrelevant for the population that is the focus of this report.

These same characteristics—along with the relative inaccuracy of the screening tests—make it unlikely that screening an asymptomatic population could improve already high survival rates. Screening test inaccuracy does, however, mean that a screening policy would produce many false-positive results (detection of noncancerous nodules), which would result in more surgical procedures than would occur absent a screening program.

The committee's review of the literature produced no randomized trials or case-control studies of thyroid cancer screening. It found only one study (Ishida and others 1988) that attempted any comparison of survival in screened and unscreened groups. That study compared Japanese women who participated in a mass screening program for breast and thyroid cancer with women who were being treated and followed for thyroid cancer in an outpatient clinic. The authors conclude that mass screening for thyroid cancer was useful because women who were screened were found to have disease that was in earlier stages than that found in the outpatient treatment group and that the screened group had a better 7-year survival rate (98 percent versus 90 percent).

Unfortunately, this study has such critical design flaws that it cannot provide a sound basis for a screening recommendation. First, the study is vulnerable to the problem of lead-time bias. Lead-time bias occurs when a screening program simply produces an earlier diagnosis and lengthens the measured period of survival after diagnosis without actually affecting true survival.

Second, people who volunteer for screening are often healthier and at lower risk than are those who do not. Compared with nonvolunteers, they would likely do better with or without screening.

Third, in this study, the screened group and the outpatient group differed in the kinds and prognosis of cancers detected, a common problem that is known as length bias. For the screened group, 99.5 percent of cancers detected were those with favorable prognosis (87.5 percent papillary, 12 percent follicular, 0 percent undifferentiated cancers). In contrast, among the cancer outpatients, 70 percent had papillary cancer, 25 percent had follicular cancer, and 4 percent had more lethal undifferentiated thyroid cancer. Seven of the 9 patients with the diagnosis of lethal undifferentiated cancer died within 5 months of the diagnosis. Thus, the screened group would be expected to have higher survival rates because of more favorable tissue type alone.

Other research cited in support of thyroid cancer screening includes two studies that focus on higher risk people—those with a history of therapeutic head and neck irradiation (Schneider and others 1985; Shimaoka and Bakri 1982). These studies report more nodules and cancers than found in the general population. A recent follow-up report on one of the study populations (a group irradiated between 1939 and 1962) concludes that nodules continue to occur in the irradiated group, but it also cautions that "thyroid ultrasound is so sensitive that great caution is needed in interpreting the results" (Schneider and others 1997).

Although it did not examine screening or screening benefits, another study (Mazzaferri and Jhiang 1994) suggests that once a thyroid cancer is manifested, delay in treatment lowers the survival rate. This study assessed a cohort of 1,355 patients with papillary or follicular cancer; the group was followed for about 25 years. Information was collected from patient records on when the patient or doctor first noticed a thyroid nodule and how long it took until surgery was undertaken. In some cases it took many years. The median time from the first recorded tumor manifestation to initial therapy was 4 months (range, <1 month to 20 years). The patients who died of cancer had a median delay of 18 months from the time a tumor was first clinically recognized (by palpation) compared with 4 months for those who survived ($p < 0.0001$, by Wilcoxon rank sum). Cancer mortality was 4 percent in patients who underwent initial therapy within a year compared with 10 percent in the others. Thirty-year cancer-specific mortality rates were 6 percent and 13 percent in the two groups, respectively. The reasons for the treatment delays are not clear and may have been related to diagnostic errors. The authors do not recommend screening for thyroid cancer but do argue for prompt assessment for larger nodules (1.5 cm or larger) and prompt treatment for those diagnosed as malignant.

Table 4.1 summarizes the committee's assessment of probabilities relevant to a screening program based on palpation and some of the consequences of such a program. It does not include probabilities or effects for thyroid cancer mortality and morbidity for lack of supporting data.

INFORMATION AND DECISIONMAKING

Explaining Potential Benefits and Harms to Patients

Whenever the issue of explaining screening risks and benefits arises, recommendations about screening strategies need to consider people's comprehension of information and explanations. Public health personnel and clinicians often rely on written materials to explain the pros and cons of screening, but literacy, more broadly, reading comprehension, limits the effectiveness of such materials. Lan-

TABLE 4.1 Summary of Evidence for Screening by Palpation for Nodules >1.5 cm

Probability of having a nodule (all age groups)	
Any size	**0.35**
<0.5 cm	**0.13**
0.5-0.9 cm	**0.10**
1.0-1.4 cm	**0.07**
1.5 cm or larger	**0.05**
Probability given a nodule that cancer is present	
Nodules smaller than 1.5 cm	**0.03**
Nodules 1.5 cm or larger	**0.10**
Probability of having cancer	
All sizes	**0.0140**
<0.5 cm	**0.0039**
0.5-0.9 cm	**0.0030**
1.0-1.4 cm	**0.0021**
1.5 cm or larger	**0.0050**
Sensitivity of palpation for nodules	
<1.0 cm	**0.00**
1.0-1.4 cm	**0.55**
1.5 cm or larger	**0.75**
Specificity of palpation for nodules	**0.95**
Fine needle aspiration (FNA)	
Sensitivity for cancer	**0.80**
False positive rate including indeterminate and unsatisfactory samples with no cancer	**0.25**
Complications of total thyroidectomy	
Recurrent laryngeal nerve injury	**0.030**
Hypoparathyroidism	**0.026**
Summary: Events per 10,000 patients screened	
Diagnosed to have nodules (true positives)	**760**
Falsely diagnosed to have nodule (false positive)	**325**
Undergo at least one FNA	**1085**
Undergo at least one FNA and have cancer	**49**
Undergo at least one FNA but do not have cancer	**1036**
Undergo lobectomy after true and false positive FNA	**402**
Cancers diagnosed	
Any size	**39**
<0.5 cm	**0**
0.5-0.9 cm	**0**
1.0-1.4 cm	**9**
1.5 cm or larger	**30**
Cancers missed	
Any size	**101**
<0.5 cm	**39**
0.5-0.9 cm	**30**
1.0-1.4 cm	**12**
1.5 cm or larger	**20**
Surgical complications	**23**

guage and comprehension problems also limit the effectiveness of oral communication. For both written and oral communication about the benefits and harms of screening, "numeracy" (a patient's ability to correctly interpret numerical information) is a particular concern.

Directly relevant research on numeracy and communication is limited. One recent study of women's interpretations of information about screening mammography concluded that "both accuracy in applying risk reduction information and numeracy were poor (one third of respondents thought that 1,000 flips of a fair coin would result in less than 300 heads)" (Schwartz and others 1997).

The same study concluded that the way information was framed was important for accurate understanding. Particularly important is information about baseline risk that allows people to look at information on screening outcomes in terms of change from baseline rather than in a vacuum. The study found best results when information was presented in terms of absolute risk compared to baseline (e.g., "reduction in risk to 4 in 1,000 with screening from 12 in 1,000 without screening"). Other research indicates that people find frequency information ("4 in 1,000") more understandable than probabilities ("0.4 percent") (Gigerenzer 1996). In addition, people tend to find a treatment more attractive when risk is expressed in terms of gain ("90 percent of patients with this treatment survive") rather than loss ("10 percent die") (Hux and Naylor 1995; Mazur and Hickam 1990b; O'Connor 1989; McNeil and others 1982). Choices about graphic displays of information likewise can affect assessments of benefits and risks (Sandman and others 1994; Mazur and Hickam 1990a; Huff 1954). Continued assessment of alternative graphic formats is important, particularly for people of limited numeracy.

Ransohoff and Harris (1997) suggest that the way mammography screening for breast cancer in younger women has been debated may reflect how numerical information is framed and perceived. For example, advocates for screening might describe 16-18 percent reductions in relative risk for death, whereas skeptics might refer to 1-2 fewer deaths per 1,000 women who have been screened annually for 10 years. Berry (cited in Ransohoff and Harris) has observed that the benefit sounds slight when it is expressed as an average of 3 days of life gained for women screened in their 40s, whereas the benefit sounds large when described as extending the lives of 800 American women in their 40s. The problems of presenting information appropriately and informing people adequately about risks and benefits are compounded because there is little direct evidence about the consequences of different strategies under different circumstances, and estimates of benefits and risks have been derived from indirect evidence of uncertain relevance.

Evidence, Decisionmaking, and Patient Preferences

Assessing the balance of benefits and harms of an intervention generally involves objective and subjective elements alike. The objective component involves the probabilities and magnitudes of potential benefits and harms, which can be evaluated through clinical or epidemiologic research. The choice of whether the benefits outweigh the harms is a personal, subjective judgment about the relative importance of potential outcomes based on individual preferences, life plans, and priorities. For example, when carefully informed about the probabilities and nature of outcomes of different clinical management strategies for localized prostate cancer, some patients will prefer not to undergo surgery, which carries some risk of incontinence or impotence; others faced with the same information will prefer surgery (Flood and others 1996; Beck and others 1994; Litwin 1994).[8]

The ways of considering patient preferences and involving patients in decisionmaking range from what might be generally characterized as, simply, good communication to formal shared decisionmaking. The latter involves a structured process in which the clinician does not uniformly encourage or discourage a treatment but instead presents the patient with the available options, reviews the potential benefits and harms associated with each, discusses the probability and magnitude of those outcomes and the quality of the evidence on which the estimates are based, helps the patient evaluate the potential importance of those outcomes in his or her life, assesses patient comprehension of the information and options discussed, and, if necessary, provides information again in ways the patient can understand (Morgan and others 1997; Woolf 1997; Kaspar and others 1992). The choice preferred by the patient is prescribed. The physician's preference or recommendation is offered if the patient inquires.

When the evidence of screening benefits or harms is limited or weak, when patient perceptions of benefits and harms are variable or not well understood, or when patient preferences about outcomes are crucial to good decisionmaking, then the strong involvement of the patient in a process of shared decisionmaking or collaboration between clinician and patient becomes particularly important (Morgan and others 1997; Ransohoff and Harris 1997; Woolf 1997; Flood and others 1996; Barry and others 1995; Emanuel and Emanuel 1992; Kasper and others 1992). In decision theory, the most important factor in choosing a course of action for such "close call" situations can be the patient's views about possible benefits and harms (Pauker and Kassirer 1997; Kassirer and Pauker 1981).

As a structured and tested process, formal shared decisionmaking is still

---

[8]Physicians sometimes have troubling misperceptions about their patients' values and preferences. For example, Leard and others (1997) gave 100 patients structured information about the benefits and harms of four screening tests for colorectal cancer. When asked which they would prefer, 38 patients chose colonoscopy, 31 chose fecal occult blood testing, 14 selected barium enema, and 13 chose sigmoidoscopy. For this group of patients, a physician who always recommended sigmoidoscopy (as some do) rather than asking these patients their preferences would subject fully over 8 out of 10 to a test other than the one they would prefer if they had been properly informed about their options.

evolving, and the effectiveness and feasibility of specific methods or techniques (e.g., computer programs, videotapes, printed materials, and formats for presenting quantitative information) for informing and deciding are still being evaluated. A formal process of shared decisionmaking is neither necessary nor appropriate in every clinical circumstance. In some medical emergencies, for example, it is not feasible. Moreover, clinicians' time with patients and patients' capacity to absorb health messages are finite, and time spent providing information about services of little or no demonstrated benefit could be time taken away from care or counseling with greater potential health benefit. Finally, many patients want a specific recommendation from their physician more than they want to play an active part in making decisions (Nease and Brooks 1995; Ende and others 1989). Still, even when formal shared decisionmaking is not appropriate and when a clinician actively encourages or discourages a particular choice, the clinician is obliged to provide a patient with information about benefits and harms and to listen to and address the patient's concerns.

RECOMMENDATIONS OF OTHERS

In addition to commissioning the literature review included in Appendix F, the committee investigated screening guidelines developed by others (Turkelson and Mitchell 1998). The review searched for screening guidelines for the general public and for people exposed to radiation. The number of thyroid-cancer guidelines located is fairly small, which is not surprising given the low prevalence and mortality of the disease.[9]

The U.S. Preventive Services Task Force guidelines (USPSTF 1996) recommend against screening for asymptomatic adults or children using either palpation or ultrasound. They characterize the evidence for this negative recommendation as fair. For asymptomatic persons exposed to upper-body radiation in infancy or childhood, they state that "recommendations for periodic palpation of the thyroid gland in such persons may be made on other grounds, including patient preference or anxiety regarding their increased risk of thyroid cancer" (USPSTF 1996). They grade this as a "C" or weak recommendation, which is defined to mean that the evidence is "poor" but that recommendations could still be made on other grounds, such as potential benefit being great and potential harm being minimal. Citing the lack of evidence, the Canadian Task Force on the Periodic

---

[9]The committee did not consider research protocols to constitute screening recommendations. For example, in research to investigate the effects of radioactive releases from the Hanford Nuclear Facility, protocols for examining exposed individuals for thyroid nodules and other conditions have been developed for the Hanford Thyroid Diseases Study (www.fhcrc.org/science/phs/htds) being conducted by the Fred Hutchinson Cancer Research Center. Such research, of course, could clarify the risks of exposure to I-131 or the outcomes of screening.

Health Examination (CTFPHE 1994) declined to make any recommendations on general or targeted screening.

The American Cancer Society (ACS 1998b) recommends thyroid palpation as part of a routine health check-up, every 3 years for those under 40 and every year thereafter. No scientific literature is cited in support of the recommendation.

A press release from the American Thyroid Association (ATA 1997) after publication of the NCI (1997a) report suggests that "people who feel that they have been exposed to significant amounts of fallout and feel that they are at particular risk might wish to see their physician for a neck examination" (p. 1). The American Association of Clinical Endocrinologists (AACE 1997b) recommends self-examination ("thyroid neck check") for thyroid nodules. It cites high prevalence of thyroid cancer as a rationale (but does not document or explain this assertion, which is not consistent with evidence cited here) but notes the disease's low mortality. It does not cite scientific evidence of efficacy or effectiveness.

The most direct recommendation for thyroid cancer screening in persons exposed to I-131 has been proposed (under the label "medical monitoring") by the Agency for Toxic Substances and Disease Registry (ATSDR 1997).[10] ATSDR recommends screening for an estimated 14,000 people, most of whom live in a limited geographic area in Washington State around the Hanford Nuclear Reservation. That group is considered to be at risk of thyroid cancer and other thyroid disease from childhood exposure (estimated at 0.1 Gy [10 rad] and above) to I-131 releases from Hanford between 1945 and 1951. Medical monitoring is defined as "periodic medical testing to screen people at significant increased risk of disease" (ATSDR 1997) The protocol recommends palpation and blood tests for most of those screened, with follow-up visits and possibly FNA biopsy or ultrasound examination for people with detected nodules 1 cm or larger. For a higher exposure subpopulation (0.25 Gy or higher), it recommends that ultrasound be "offered" at the initial visit (p. 37). In making its recommendation for medical monitoring, ATSDR states "the evidence supports benefits associated with thyroid screening for populations exposed to radiation" (p. 63). In particular, as evidence of the survival benefits of early detection, it cites the study of Ishida and others (1988), which was critiqued earlier in this chapter. In response to critiques

---

[10]ATSDR is responsible, under section 104(i)(10) of the Comprehensive Environmental Response, Compensation, and Liability Act of 1980 (commonly called the Superfund act), for initiating health surveillance programs for "populations at significant increased risk of adverse health effects as a result of exposure to hazardous substances." A major goal of ATSDR is to expand the knowledge base about such effects, but the medical-monitoring program is not designed as a research program (e.g., participants would be self-selected). ATSDR's very preliminary estimate of first-year costs for its recommended program was $9.5 million for 14,000 eligible people (0.1 Gy or more exposure in childhood) with an assumed participation rate of 100 percent spread evenly across the first 2 years. That amounts to about $680 per eligible person, exclusive of treatment costs for those referred for surgery and any direct (e.g., transportation) or indirect (e.g., lost work time) costs incurred by those being screened.

from reviewers of a draft of its report, ATSDR stated that the Japanese study provided "important evidence that screening results in improved survival" (p. 113).

This committee paid careful attention to the ATSDR analysis and recommendation because it involves a population historically exposed to varying levels of I-131. This committee did not agree with ATSDR that systematic thyroid screening should be recommended for asymptomatic people whether or not they had exposure to I-131. In particular, this committee concluded that research did not support systematic screening and that the Ishida study cited by ATSDR was seriously flawed and did not provide valid, usable evidence of benefit. DHHS will need to establish some process for managing or resolving this conflict. If political pressure prompts DHHS to decide to recommend or encourage screening, it should make clear that scientific evidence does not support the recommendation.

In addition to considering specific recommendations about thyroid cancer screening from other groups, the committee also examined the approach to screening taken by one large managed-care organization, Group Health Cooperative (GHC, formerly Group Health Cooperative of Puget Sound, now part of Kaiser/Group Health) (Thompson 1996). GHC's preventive services program, which actively promotes preventive services of demonstrated value, does not explicitly include thyroid cancer screening. However, it is relevant here to examine the program's approach to other tests and conditions for which evidence to support screening is insufficient. In information for patients and clinicians, the organization will state when there is insufficient evidence to support routine screening for a problem. Rather than preclude screening, GHC may recommend shared patient-clinician decisionmaking about screening. In addition, GHC also may employ various means (including an educational program that uses opinion leaders, online guidelines, and supporting documentation and references) to inform its clinicians about the evidence on test accuracy and on the lack of research showing benefit from early detection through screening.

COMMITTEE FINDINGS AND RECOMMENDATIONS

Findings

When evaluated against the criteria for screening recommendations set forth earlier in this chapter, the evidence reviewed by the committee does not support a clinical recommendation for routine screening for thyroid cancer in asymptomatic persons exposed to radioactive iodine from nuclear weapons testing at the Nevada Test Site.

First, thyroid cancer is rare in the general population. By age 95, it is estimated that about 1 in 400 men and 1 in 150 women might be diagnosed with some form of thyroid cancer. In contrast, about 1 in 8 women might be diagnosed with breast cancer and 1 in 5 men with prostate cancer.

Second, exposure to I-131 in childhood does appear to increase the risk of thyroid cancer. Exposure to 0.1 Gy (10 rad) of I-131 at the age of less than 1 year has been estimated to yield a lifetime risk of thyroid cancer of about 1 in 240 for men and 1 in 90 for women.

Third, for most people, there will not be enough information to identify accurately the level of exposure to I-131 from the Nevada nuclear tests.

Fourth, papillary thyroid cancer, the most common form of naturally occurring thyroid cancer and the form linked to radiation exposures, can be effectively treated surgically and has a high survival rate, regardless of cause, when detected by routine clinical practice *without screening*. The 10-year cancer-specific survival rate for persons with papillary carcinoma is estimated at 95 percent and the 30-year survival rate is estimated at 90 percent.

Fifth, there is no evidence that early detection of thyroid cancer through systematic screening (rather than through routine clinical care) improves health outcomes or has benefits that outweigh harms. It is still possible—given the lack of directly relevant research—that early detection through routine screening might offer some net benefit.

Sixth, routine screening for thyroid cancer by palpation and, especially, by ultrasound will identify many nodules, most of which will not be malignant.

Seventh, FNA biopsies will find cancer in a few nodules, will not find cancer in most nodules, and will yield a significant proportion of indeterminate or unsatisfactory samples (20 to 30 percent or more) that may lead to unnecessary surgery for many people who do not have thyroid cancer or who have very small cancers that would never progress to cause health problems.

Overall, the committee found that routine screening for thyroid cancer does not meet the criteria set forth earlier in this chapter (Figure 4.1) related to prevalence, accuracy of screening tests, or evidence of improved survival or benefits that exceed harms.

Recommendations

Public Health and Clinical Policies

The committee recommends against public programs and clinical policies to promote or encourage routine screening for thyroid cancer in asymptomatic people possibly exposed to radioactive iodine from fallout as a consequence of the nuclear tests in Nevada during the 1950s.

The lack of evidence that early detection of thyroid cancer through routine screening of asymptomatic persons improves health precludes a positive recommendation to screen people routinely for a disease characterized by slow progression, high survival rates without screening, and high rates of false-positive test results that can lead to unnecessary surgery and other harms, including some, such as anxiety and insurability problems, that are not well understood. The committee's recommendation is counter to ATSDR's recommendation in favor of medical monitoring for a subset of people exposed to releases of radioactive iodine from the Hanford Nuclear Reservation in Washington State in the 1940s. The committee recognizes that conflicting policy recommendations can be confusing to the public and distressing to some, but it firmly believes that the scientific evidence supports its position. DHHS will obviously require some process for managing or resolving this conflict.

It is appropriate for a clinician who sees a concerned patient to discuss that patient's concerns and history and decide jointly about screening.

Given popular fears of cancer and concern about radiation, the often modest reach of public information programs, and conflicting recommendations from other groups, clinicians will likely see some patients who express concern about possible exposure to radioactive fallout and who request screening for thyroid cancer.[11] Although the committee recommends against policies that encourage or promote routine screening, it is essential that clinicians respond sensitively and constructively to concerned patients who come to them seeking advice. Such a response will involve listening to the patients' concerns; discussing their possible exposure to iodine-131 and other risk factors for thyroid cancer; explaining that thyroid cancer is uncommon even in people with some exposure to I-131 and that the thyroid cancer linked to I-131 exposure is rarely life threatening; describing the process, benefits, and harms of screening and the lack of evidence showing that people are better off with it than without it; checking patient understanding of the information presented; and jointly deciding how to proceed. If a nodule is palpated, FNA biopsy would normally be proposed, again with a discussion of what is known about its accuracy, benefits, and harms and about the patient's prospects without further testing.

Once the committee concluded that it could not recommend a routine program of screening, three points led the committee away from an explicit recommendation that physicians forego screening and toward the recommendation for patient-physician discussion and decisionmaking. The committee's recommendation that patients and physicians share in decisionmaking about thyroid cancer screening rests on several points. First, the evidence of the effects of systematic screening is not negative; it is absent. Neither positive nor negative effects have been documented from soundly designed clinical research. Second, understand-

---

[11]Such testing is not screening in conventional public health terms, which usually assume that "the initiative for screening . . . comes from the investigator or agency providing care rather than from a patient" (Last's Dictionary of Epidemiology, 1988, pp. 118-119). However, because such testing makes use of the same procedures and because one goal of individuals who request the tests is early detection of disease, the committee decided to continue to use the term "screening" to describe testing requested by patients.

ing of the benefits and harms of screening as perceived by patients—especially those at higher risk—is quite limited, although some committee members believe the potential harms exceeded the potential benefits. Certainly, the committee heard about the anxieties of those who had become aware of their possible exposure to I-131 from the Hanford facility and who were concerned about additional exposure from the Nevada tests. Third, small nodules (<1.5 cm) are much less likely to be detected by palpation than by ultrasound. Because small nodules seldom cause health problems, palpation has less potential than ultrasound to cause harm (in the form of unnecessary follow-up tests and surgery), which is important when the benefits of routinely screening asymptomatic people are uncertain (Sox 1998).

The communication and decision strategy described above is less formal than some strategies of shared decisionmaking described earlier in this chapter. The rationale is primarily practical. In the committee's view, it is feasible and desirable to listen to a patient's concerns about possible radiation exposure and its consequences and to discuss possible benefits and harms of different courses of care. It is less feasible for a clinician faced with a concerned patient to postpone a safe, noninvasive, personal examination while the patient reads a brochure, views a videotape, or otherwise is involved in a structured education and decisionmaking process. (If ATSDR proceeds with its medical monitoring program, it might nonetheless consider testing more and less formal ways of discussing benefits and harms with patients and reaching decisions about screening.) In addition to these practical considerations, a number of committee members also believed that formal shared decisionmaking is most appropriate for certain kinds of "close call" situations as explained above and that thyroid cancer screening does not qualify as such a situation.

Few physicians outside areas where radiation risk has been a prominent issue locally will be prepared to discuss radiation exposure and cancer risk. They will need easy access to information that will prepare them to counsel concerned patients. A sample information sheet for physicians is presented later in this chapter.

For asymptomatic patients concerned about exposure to I-131 and thyroid cancer, the committee recommends against using ultrasound examination for screening either initially or following a negative result from palpation.

The committee's recommendation against screening by ultrasound examination for patients who come to their physicians concerned about I-131 exposure is based on the points noted earlier, in particular, the lack of evidence of screening benefit and the probability that ultrasound will reveal many small nodules that are unlikely to cause harm. It was also based on the additional point that if no nodule is detected by palpation, the likelihood is substantially diminished that an ultrasound will identify a nodule with microcancers that are unlikely to progress to

TABLE 4.2 Bayesian Analysis of Screening of Nodules ≥1.5 cm by Palpation

Test conditions for nodule detection:
 Sensitivity of test, 75%
 Specificity of test, 95%

Patient Status	Prior Probability of a Nodule	Probability of Negative Exam	Product of Probabilities	Revised Probability of a Nodule
Nodule 1.5 cm or larger	5%	25%	125	1.37%
No nodule 1.5 cm	95%	95%	9,025	98.63%
			$\Sigma = 9,150$	

cause problems. The magnitude of this effect can be calculated by applying Bayes' rules (see Addendum 4A) (Pauker and Kopelman 1992; Pauker and Kassirer 1987). Table 4.2 illustrates the effect based on assumptions that the prevalence of nodules of 1.5 cm and larger is about 5 percent, that the sensitivity of physical examination in identifying nodules of this size is 75 percent (meaning that the chance of no such nodule being found when one is actually present is about 25 percent), and that the specificity of the exam is 95 percent (meaning that the chance of a clinician falsely feeling such a nodule when none exists is 5 percent). Application of Bayes' rules, as shown below, reveals that the likelihood of a nodule—given that none was palpated—is now less than 1.4 percent rather than 5 percent. The likelihood of no nodule of this size—given that none was palpated— is 98 percent rather than 95 percent. If 10 percent of nodules of this size are cancerous and if no nodule is detected through palpation, then the likelihood of a cancerous nodule larger than 1.5 cm is 0.137 percent (0.10×0.0137) meaning one such cancer would be detected for every 730 people screened by ultrasound after a negative physical examination. (Because clinical studies give varying estimates of the variables used in the table and because experts may disagree on the appropriate assumptions, the table can be recalculated using other numbers to see how probabilities change. See also the addendum to this chapter.)

Information and Communication

 The committee recommends that DHHS develop a program of information and education for the concerned public and clinicians that builds on the analyses, conclusions, and information approaches described in this chapter and the experience the Department and others have gained in developing similar informational materials.

 Patient and Public Information Although screening programs—whether or not they are supported by scientific evidence—can be a popular response to concerns about cancer risk, the U. S. Department of Health and Human Services (DHHS) will best serve the public by declining to promote routine screening for those exposed to I-131 from the Nevada tests. The NCI has provided written and electronic information about its report and possible health problems, but this information should be reevaluated, refined, and expanded based on experience gained in the last year and on the experience of others in communicating similar kinds of information. Others with relevant experience in communicating about radiation risk include the Centers for Disease Control and Prevention, the Hanford Health Information Network, and the Hanford Thyroid Disease Study. Although managed care organizations are unlikely to have much experience with communicating about radiation risk, organizations such as Group Health Cooperative and Stanford's Center for Patient Preferences may have useful insights about developing effective public information about screening tests. Chapter 5 discusses the importance of early public involvement in developing such an information and communication program.

Also as discussed in the next chapter, the Department's information program should include several kinds of materials including brochures or the equivalent that clinicians, public health departments, managed care plans, and others could give to concerned individuals. Such materials should explain basic facts about the Nevada tests, I-131 and thyroid cancer (including symptoms) and provide general education about screening tests, including explanations of such concepts as false-positive and false-negative test results and their consequences. An Internet site is useful but should not be relied on to the near-exclusion of other kinds of information.

Particularly in materials designed for patients and the public, the limited research on the communication of information about health risks and benefits suggests that the risk of thyroid cancer related to exposure to I-131 from nuclear-bomb testing should be described in both quantitative and qualitative terms. The quantitative information for patients should include an estimate of baseline risk and should be expressed as frequencies (rather than probabilities) and in absolute rather than relative terms. Although some information should be written specifically for the lay public, it should include clear links to more extensive information aimed at clinicians. Box 4.1 presents a sample information page for physicians to provide patients. It is constructed to be consistent with the limited research cited in this report.

Information materials developed by NCI should be tested to assess whether likely audiences interpret the materials accurately. Professional groups and health care organizations may also want to develop and test both written information and sample scripts to guide face-to-face discussions between clinicians and patients. It is particularly important for such patient information to accompany any materials that provide people with a method or model for calculating their exposure to I-131 (e.g., like the method presented on the NCI Web site). Whatever information NCI develops, it should also be tested with advocacy groups and

BOX 4.1
Sample Information about Thyroid Cancer,
Iodine-131, and Screening

Thyroid cancer is rare and usually not life-threatening. Approximately one-third to one-half of Americans between 40 and 60 have small lumps or nodules in their thyroid glands, most of which are too small to be felt. Possibly 10 to 15 percent of people in this age group will have nodules that contain tiny cancers, most of which grow very slowly and will not progress to cause problems. For the most common kind of thyroid cancer, papillary cancer, about 99 out of 100 people with this diagnosis will be alive 5 years after diagnosis and about 90 of 100 will be alive 30 years later.

Nuclear weapons tests in Nevada in the 1950s and 1960s exposed many Americans to small doses of radioactive iodine. Although radioactive iodine sometimes causes papillary thyroid cancer, the risk of thyroid problems for most potentially exposed people is still small. From birth to death, an average American woman has about 1 chance in 150 of being diagnosed with thyroid cancer without exposure to radioactive iodine and about 1 chance in 90 following exposure as a child under one year of age. For the average man, the probability of developing thyroid cancer is about 1 in 400 without exposure and 1 in 250 with exposure. Regardless of how they got thyroid cancer, most people will not die from the disease.

It is usually not possible to identify accurately how much radiation someone received from the Nevada weapons tests. For this and other reasons, it is also difficult to estimate how likely it is that a particular person will develop radiation related thyroid cancer. In general, people are more at risk if they were under the age of 10 years and also drank milk (especially milk from "backyard" cows and goats) in the years 1951-1962.

Routine screening for thyroid cancer or other thyroid disease in people who are unaware of any problems (e.g., a lump they can see or feel) is not recommended, whether or not people have been exposed to I-131. This is because the available tests are often inaccurate and can lead to unnecessary testing and treatment and because screening has not been shown to improve health outcomes. People concerned about thyroid cancer should make a decision about screening in consultation with their doctor following discussion about the accuracy of the screening tests, the possible benefits and harms of screening, and the lack of evidence that thyroid cancer screening saves lives.

with representative people in areas where information related to the NCI report was covered in the news media.

Information for Physicians For a decision about screening to be an informed one, clinicians will need to explain carefully the possible benefits and harms of screening for thyroid cancer. Although educational materials for physicians and materials for them to give their patients can help prepare physicians for discussions with patients, actual discussions between patient and physician cannot be governed by a single script but will need to be adapted to some degree to the characteristics and needs of specific patients. What is suitable for a high school physics teacher will likely differ from what is helpful for someone with less than a high school education. In any case, it is appropriate for clinicians to check patient comprehension of oral information and instructions.

In addition to patient and public information, NCI should provide information explicitly designed for clinicians. Except in some areas in the West, most clinicians probably have not seen and would not now expect to see patients concerned about I-131 exposure and cancer risk. This may change as a result of news coverage surrounding publication of this report. Clear, easily available information from NCI could be particularly helpful to busy, relatively uninformed physicians facing new inquiries about I-131 and thyroid cancer.

Although the National Cancer Institute Web site already includes information about its report on I-131 fallout and about thyroid cancer, it does not have information directed explicitly to physicians. When NCI reviews its Internet strategy, information for physicians should be one priority. A sample information sheet for physicians might read as follows:

IODINE-131 AND THYROID CANCER INFORMATION
FOR PHYSICIANS [SAMPLE]

News coverage about thyroid cancer and fallout from the 1950s nuclear tests in Nevada may prompt people concerned about possible radiation exposure to seek advice from physicians. The points below briefly summarize information that may help physicians counsel their patients. Additional information is available from the report of the National Cancer Institute, "Estimated Exposures and Thyroid Doses Received by the American People from Iodine-131 in Fallout Following Nevada Atmospheric Nuclear Bomb Tests" (NIH Pub. No. 97-4264, 1997), the Institute of Medicine/National Research Council report "Exposure of the American People to Iodine-131 From Nevada Nuclear-Bomb Tests: Review of the National Cancer Institute Report and Public Health Implications," and several Web pages at the National Cancer Institute site (http://rex.nci.nih.gov).
In brief:

• The risk of thyroid cancer may be higher than average in people with childhood exposure to iodine-131 (I-131) fallout from nuclear weapons testing in Nevada from 1951 to 1962. For most of this group, the risk of developing thyroid cancer is still small—from birth to death, less than 1 chance in 90 for women and 1 in 240 for men. This is less than many other cancers in the general population.

• Thyroid cancer is rarely life-threatening; 30-year survival is over 90 percent for papillary thyroid cancer, the form linked to radiation exposure.

• Accurately identifying people's past exposure to I-131 is usually not possible because necessary data on the key risk factors from four decades ago are generally not available or are unreliable.

• Routine screening for thyroid cancer is not recommended because the available tests produce a large number of false-positive results, there is no evidence that early detection by screening improves outcomes, and the benefits of screening may not outweigh the harms.

Why may my patients be concerned? News media have covered two recent reports on exposure to radioactive iodine in fallout from about 100 above-ground, nuclear weapons tested in Nevada in the 1950s and 1960s. These stories have reported that this exposure is linked to thyroid cancers.

The first report, mandated by Congress and published in 1997 by the National Cancer Institute, provided county-level estimates of the American people's exposure to I-131. A related memo estimated that an 11,000 to 212,000 excess cases of thyroid cancer were likely caused by the I-131 exposure but epidemiological analyses suggest the number is probably in the lower part of this range. An Internet site provides the public a complicated method to estimate individual exposure (http://rex.nci.nih.gov).

In 1999, the Institute of Medicine and the National Academy of Sciences, as requested by the U.S. Department of Health and Human Services, published a report assessing the health implications of I-131 exposure from the Nevada tests and advising the government on appropriate responses. The IOM/NRC report concluded that the available data from events four decades past were insufficient to permit either reliable or valid county- or individual-level assessments of exposure to I-131. It also concluded that there was no evidence to determine whether screening for thyroid cancer would improve survival or other health outcomes, taking possible harms of mass screening of the population into account.

Does exposure to I-131 cause cancer? Although there is still some disagreement, it is now generally accepted that exposure to I-131 at young ages— especially under age 10—can cause thyroid cancer. The strongest evidence comes from the studies of thyroid cancer in children exposed from the nuclear accident at Chernobyl in 1986. Other suggestive evidence links thyroid cancer to upper body external radiation during childhood.

The magnitude by which risk is increased is small, however—probably less than the normal variation in thyroid cancer in different populations around the world. In the general U.S. population, the chance of being diagnosed with thyroid cancer by the age of 95 is roughly 1 chance in 400 for men and 1 in 150 for women. For someone who received 0.1 Gy (10 rad) of I-131 as a child under age 1, the lifetime risk might be roughly 1 chance in 240 for men and 1 in 90 for women. As noted earlier, papillary thyroid cancer, the form related to radiation exposure, is rarely life threatening. By way of comparison, women have about a

1 in 8 lifetime risk to age 95 of being diagnosed with breast cancer, and men have about a 1 in 5 risk of being diagnosed with prostate cancer.

For those diagnosed with papillary carcinoma, the risk of dying from thyroid cancer is very small. Ten-year survival rates are about 95 percent and 30-year survival is about 90 percent. For the population in general, the lifetime risk over 95 years of dying of thyroid cancer (all forms) is considerably less than 1 in 1,000. In comparison, women have about a 1 in 30 risk of dying of breast cancer and men have about the same risk of dying of prostate cancer.

Is thyroid cancer linked to iodine-131 different from naturally occurring thyroid cancer? The cancer linked to I-131 exposure is papillary carcinoma, which has very high survival rates (95 percent), is a usually less aggressive type of thyroid cancer, and has a very good prognosis without screening (95 percent survival at 10 years, 90 percent at 30 years). For the population at risk, today's middle aged adults, there is no evidence that cancer related to childhood radiation exposure differs from naturally occurring cancer.

Can I identify whether a patient is at high risk of thyroid cancer from iodine-131 exposure? Unfortunately, in most cases the answer is no—except that there is virtually no risk for someone who was an adult during the testing period. For those who were children at the time of the testing, one problem in estimating exposure is that fallout data were collected in very few places in the United States. This information is not sufficient to make good estimates about fallout in places where data were not collected. Also, because the primary way that I-131 gets to the thyroid is through drinking milk, you would need to know how much milk a person drank as a young child, whether it came from a cow or a goat (because goats' milk concentrates I-131 more effectively than cows' milk), and whether the milk came from a backyard cow or other source that left little time for natural decay of radioactivity. Research indicates that dietary recall data are highly unreliable, especially four decades after the fact. In addition to the uncertainties in estimating I-131 exposure, there are also uncertainties involved in estimating the probability of thyroid cancer related to particular levels of exposure, especially at the low levels generally linked to the Nevada weapons tests.

Should people be screened for thyroid cancer? An Institute of Medicine committee made up of physicians (including specialists in endocrinology, radiology, pathology, surgery, and family practice) and others with expertise in public health and evidence-based medicine found that there is insufficient evidence to recommend routine screening for thyroid cancer or other thyroid disease in people who are asymptomatic, whether or not they have been exposed to I-131. For physicians who see patients concerned about thyroid cancer and interested in screening, a process of shared decisionmaking is appropriate. The U.S. Preventive Services Task Force reached similar conclusions about screening for thyroid cancer in the general population. The reasons for this policy are based on concerns about benefits and harms of screening, outlined below, and do not relate to the costs of a screening program.

The problem with screening for thyroid cancer is that palpation is not a very reliable test for thyroid nodules or thyroid cancer. Ultrasound is more accurate in finding nodules, but because it can detect very small lesions, it will find nodules in perhaps 30 to 50 percent of older adults. The great majority will be benign, but ultrasound can't distinguish them. Many people will then be referred for FNA biopsy, the results of which are also not perfect. If the physical examination is normal and no symptoms are reported, ultrasound is not recommended.

One potential result of a screening program is that many people would have surgery when they did not have cancer or had cancers that would never cause problems. This extra surgery would be acceptable if the added vigilance from screening meant better survival and reduced morbidity from thyroid cancer. Survival rates for papillary cancer are already high, however, and no evidence exists to determine whether screening will produce better results. This combination of unproved benefits and clear possible harms is why a systematic screening for thyroid cancer is not recommended.

What should I tell patients interested in testing for thyroid cancer? Given the tradeoff of benefits and harms associated with screening, the choice of whether to proceed with screening is a personal decision that should be made, whenever possible, with the input of the patient. If you have a patient who is interested in testing, it is important to advise the patient of the potential benefits and harms so that he or she makes an informed decision about whether screening is worthwhile. Some of the quantitative information presented above may help in your discussions with patients, but people often don't interpret quantitative information correctly. It is appropriate to check their understanding.

You might start by explaining that thyroid cancer is uncommon and usually not life-threatening. Possibly 10 percent of the U.S. population may have tiny thyroid cancers, few of which will ever progress to cause problems. You can then explain that the primary screening test is palpation, which looks for thyroid nodules using a physical examination of the neck, combined with questions about possible symptoms (e.g., hoarseness) and risk factors (e.g., radiation therapy at a young age). Possibly one-third to one-half of American adults have thyroid nodules, most of which, however, are too small to be felt. If a lump is found, you may then recommend a biopsy, which uses a thin needle to draw tissue from the lump for examination under a microscope. Most of the time they will not find cancer.

Patients should know that screening has the potential to produce harm because the tests used are not always accurate. The tests often identify questionable lumps or cells that lead people to have major surgery either when they don't have cancer or when they have a cancer that will never progress and cause problems. The surgery has very low risk of death but it does cause discomfort and scarring, will require lifelong thyroid hormone replacement therapy, and for a small proportion of cases can disrupt calcium metabolism or cause serious, permanent hoarseness.

Patients should also know that there is no scientific evidence that they will enjoy better health as a result of screening or early treatment. If, after learning of these tradeoffs, the patient still wants to be screened, palpation of the neck combined with questions about possible symptoms (e.g., hoarseness) and risk factors (e.g., radiation therapy) may be reasonable and reassuring. Normally, if a nodule is palpated, FNA biopsy is recommended. If the physical examination is normal and no symptoms are reported, ultrasound is not recommended.

Consistent with the recommendations for patient and public information, information for physicians should be tested before distribution. DHHS should enlist physician organizations in the process of developing information, testing it with representative generalists and specialists, and disseminating it. The committee recognizes that the availability of information and recommendations does not translate automatically or easily into clinical practice. Physicians facing questions from their patients will, however, have some motivation to seek guidance on iodine-131 exposure and clinical practice. Following publication of this report, DHHS could organize briefings for relevant professional organizations in an effort to enlist their support and assistance in reaching their members, for example, through journal articles, stories in professional newsletters, and presentations at national and regional meetings. If DHHS pursues the regional information development and distribution strategy described in Chapter 5, physician leaders could be involved in the planning process.

Research and Surveillance

Given the lack of research on how people understand and perceive the risks of different cancers and the benefits and harms of cancer screening, the committee suggests that DHHS consider studies to develop more knowledge about perceptions of cancer risk and screening benefit and harms and about the ways perceptions may differ depending on disease characteristics (e.g., prevalence, risk factors, mortality and morbidity, screening test accuracy, treatment consequences). Such research would be helpful in developing strategies for informing and counseling people about risks and options, for example, in deciding whether and how strategies might need to vary depending on disease characteristics. Also helpful would be research on how perceptions are affected by different ways of presenting quantitative information and different ways of structuring clinician-patient communication.

With respect to thyroid cancer or radiation-exposed populations specifically, the committee suggests that the evaluation component of the ATSDR medical monitoring program might consider the feasibility of a controlled study to compare the responses to different information formats or different counseling strategies for eligible patients who come in for screening. Although the program might contribute to knowledge in this area, the ATSDR medical-monitoring program is not likely to produce useful information about mortality effects of screening for thyroid cancer because the population that requests screening and meets eligibility criteria will be self-selected and because high long-term survival rates can be expected without screening.

The committee already has noted that the results of the Hanford Thyroid Disease Study (which is a research effort rather than a screening program) may help clarify the extent to which childhood exposure to low doses of I-131 is linked to thyroid problems in adults. This clarification should, in turn, indicate whether systematic examination of the benefits and harms of screening for nonmalignant thyroid disease in radiation-exposed persons is warranted.

In addition to supporting some additional research to inform clinical practice and public health policy, the committee also suggests that DHHS strengthen its surveillance capacity by considering ways to work closely with states to improve the quality and scope of SEER data collection and reporting. It should consider whether collection of additional patient information on residence and personal history is warranted and should review the information collected and the way information is coded to be sure that FNA and open biopsy results are being adequately captured, and that operative results and treatment of confirmed thyroid cancers are being documented.

CONCLUSIONS

This chapter has examined the potential benefits and harms of screening for people who could be at higher-than-general risk for thyroid cancer as a result of exposure to I-131 fallout. It briefly discussed screening for other thyroid disorders. The IOM committee that prepared this chapter recognized the significant uncertainties that surround the issues of I-131 exposure and cancer risk as illustrated in Figure 4.2, which summarizes the causal pathway from I-131 release to diagnosis of cancer and gives examples of the sources of variation and uncertainty associated with each step in the pathway. The IOM committee

(1) accepted the analysis presented elsewhere in this report indicating that those most exposed to I-131 from fallout were at increased risk of thyroid cancer, although much uncertainty surrounds estimates of exposure for specific individuals;

(2) concluded that the evidence did not support a positive recommendation for a program to promote systematic thyroid cancer screening for those potentially exposed to I-131 from the Nevada atomic bomb testing program;

(3) described a simplified process of shared decisionmaking about screening by palpation as a reasonable approach for people who come to clinicians with requests for screening and with concerns about their risk of thyroid cancer due to I-131 exposure;

(4) recommended against screening by ultrasound examination for those who choose screening;

(5) suggested that DHHS develop materials for physicians and for physicians to give to concerned patients as part of its program of information and education for the concerned public and clinicians; and

(6) proposed directions for research, including the development of more information about how people perceive the benefits and harms of screening and about how different ways of presenting risk information and structuring decisionmaking affect patient perceptions and understanding.

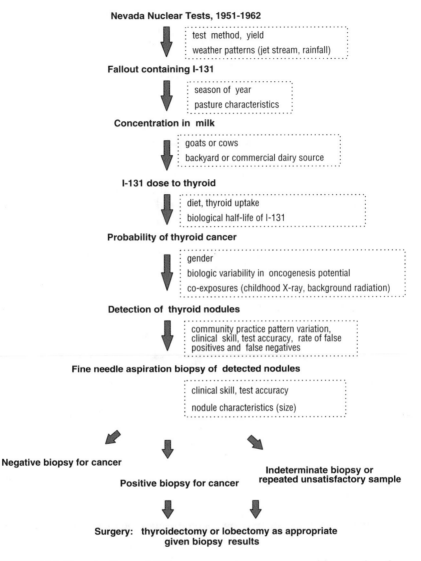

FIGURE 4.2 Causal pathway: I-131 exposure and consequences with examples of sources of variation and uncertainty.

The committee was not charged with doing a cost-effectiveness analysis, and it based its conclusions solely on clinical and epidemiologic grounds. It would not, in any case, have proceeded from analysis of effectiveness to analysis of cost-effectiveness because it found no evidence showing that screening for thyroid cancer is effective.

Although screening for those at increased risk of thyroid cancer might seem an obvious strategy, screening asymptomatic persons has value only when earlier detection of disease results in earlier treatment that improves outcomes for the screened population and when the benefits of screening exceed the harms. For thyroid cancer screening, there is no evidence of improved outcomes or of benefits that exceed harms. The committee recognizes that it will be challenging to communicate these conclusions in ways that respond to understandable public concerns, beliefs, and values and that accommodate limits on popular comprehension of quantitative information and risk analyses. The next chapter discusses how DHHS might confront this challenge.

Figure 4.2 traces the steps involved in the causal pathway connecting atmospheric nuclear weapons tests to the detection and treatment of radiation-induced thyroid cancers. For each step of the pathway, the enclosed boxes contain information about factors likely to introduce uncertainty and variation. The cumulative effect is to dramatically reduce the ability to identify those individuals at highest risk for radiation-induced thyroid cancer and to assure an individual entering the screening portion of the pathway that benefits will exceed harms.

ADDENDUM 4A: INTERPRETING SENSITIVITY AND SPECIFICITY

Screening tests are accurate when they correctly identify disease in people who have the disease (true positive) or when they correctly identify no disease in people who have no disease (true negative). Tests are inaccurate when people without the disease have a positive test result (false positive) or when people with the disease have a negative result (false negative).

Measures of sensitivity, specificity, and positive predictive value are used to assess the accuracy and efficiency of screening tests in identifying people with and without disease (Table 4.3). These measures are defined for the straightforward screening situation when the alternatives are, first, that the disease is either present or absent and, second, that the test results are either positive or negative. When indeterminate test results and conditions are factored in, the computations of sensitivity and specificity are not defined, but grouping can be done. Treating indeterminate results as positive lowers the positive predictive value of the test.

Table 4.4 highlights the importance of disease prevalence in assessing the value of screening and shows a worked example using the concepts defined above. Given the same test accuracy, the positive predictive values of the test—or the probability of disease given a positive test result—goes from 8.3 percent to 0.9 percent when the prevalence of a disease drops from 1 percent to 0.1 percent. Even for a test that is reasonably sensitive and specific, the lower the prevalence of the disease, the more false-positive results will be generated relative to true-positive results.

Table 4.5 uses the same example but presents it as a Bayesian analysis. In the column headed "revised probability," the first number (8.3 percent) is the prob-

TABLE 4.3 Definition of Terms

Term	Definition	Formula[a]
Sensitivity	Proportion of persons with condition who test positive	$a/(a+c)$
Specificity	Proportion of persons without condition who test negative	$d/(b+d)$
Positive predictive value	Proportion of persons with positive test who have a condition	$a/(a+b)$
Negative predictive value	Proportion of persons with negative test who do not have a condition	$d/(c+d)$

[a] Explanation of symbols:

	Condition present	Condition absent	
Positive test	a	b	a+b
Negative test	c	d	c+d
	a+c	b+d	

Legend:
a = true positive
b = false positive
c = false negative
d = true negative

SOURCE: USPSTF, 1996.

TABLE 4.4 Importance of Disease Prevalence

Testing Conditions:
    Size of Population = 100,000
    Sensitivity of test = 90%
    Specificity of test = 90%

If disease prevalence = 1%

	Disease Present	Disease Absent	
Positive Test	900 (a)	9,900 (b)	10,800 (a+b)
Negative Test	100 (c)	89,100 (d)	89,200 (c+d)
Probability of disease given a positive test result = 8.3% $(a / a+b) \times 100$	1000 (a+c)	99,000 (b+d)	100,000 (a+b+c+d)

If disease prevalence = 0.1%

	Disease Present	Disease Absent	
Positive Test	90 (a)	9,990 (b)	10,080 (a+b)
Negative Test	10 (c)	89,910 (d)	89,920 (c+d)
Probability of disease given a positive test result = 0.9%	100 (a+c)	99,900 (b+d)	100,000 (a+b+c+d)

TABLE 4.5 Importance of Disease Prevalence: Bayesian Probability Analysis

Test conditions:
  Sensitivity of test, 90%
  Specificity of test, 90%

Patient Status	Prior probability	Probability of positive exam	Product of probabilities ($col_1 \times col_2$)	Revised probability $col_i / \Sigma col_i$
If prevalence is 1%				
Disease	1%	90%	90	8.3%
No Disease	99%	10%	990	91.7%
			$\Sigma = 1,080$	
If prevalence is 0.1%				
Disease	0.1%	90%	0.009	0.9%
No Disease	99.9%	10%	0.999	99.1%
			$\Sigma = 1.008$	

ability of disease given a positive test and is the same as in the first part of Table 4.4. The first sum at the bottom of the column headed "product of probabilities" (10.80 percent) multiplied by 100,000 would give the number of positive tests as shown at the right side of the first part of Table 4.4 above.

ADDENDUM 4B: INTERPRETATION OF INDETERMINATE AND UNSATISFACTORY FNA SAMPLES

The terms sensitivity and specificity are clearly defined only when the test result is either positive or negative. In the case of FNA, the test results can also be indeterminate and unsatisfactory. As shown in Table 2a and 2b in Appendix F, the test performance of FNA will depend upon whether one treats indeterminate and unsatisfactory test results as positive or negative. The following analysis illustrates how treatment of these results affects the calculation of the probability of cancer. For these illustrative purposes, the analysis relies on experience from a major academic center as reported in the table below, which presents the approximate probability of positive, negative, indeterminate, and unsatisfactory results conditioned on the actual findings for the nodule aspirated.

	Cancer	No Cancer
Positive	85%	3%
Negative	7%	65%
Indeterminate	4%	20%
Unsatisfactory	4%	12%
	100%	100%

If it is assumed that the prevalence of cancer among nodules biopsied is 3 percent, then (based on the tabular algorithm for Bayes rule presented in Table 4.4) one can revise the probability of cancer based on a positive FNA result:

Positive

Diagnosis	Prior	Conditional	Product	Revised
Cancer	3%	85%	255	46.7%
No cancer	97%	3%	291	53.3%
			$\Sigma = 546$	

The probability of a positive result increases from 3 percent to 46.7 percent. Similarly, after a negative FNA result:

Negative

Diagnosis	Prior	Conditional	Product	Revised
Cancer	3%	7%	21	0.3%
No cancer	97%	65%	6,305	99.7%
			$\Sigma = 6,326$	

The probability of cancer given a negative test decreases from 3 percent to 0.3 percent.

Now consider the information provided by an indeterminate or an unsatisfactory FNA result.

Indeterminate

Diagnosis	Prior	Conditional	Product	Revised
Cancer	3%	4%	12	0.6%
No cancer	97%	20%	1,940	99.4%
			$\Sigma = 1,952$	

Unsatisfactory

Diagnosis	Prior	Conditional	Product	Revised
Cancer	3%	4%	12	1.0%
No cancer	97%	12%	1,164	99.0%
			$\Sigma = 1,176$	

In both cases, the probability of cancer decreases (from 3 percent to 0.6 percent and 1 percent, respectively). This change occurs because both of these results carry information: The chance of either an indeterminate or an unsatisfactory result is only 4 percent if the nodule contains cancer but is higher (20 percent and 12 percent, respectively) if the nodule is free of cancer, because cancers tend to be more cellular and because criteria for the diagnosis of cancer are often more explicit than criteria for diagnosing benign disease.

How, then, should an indeterminate result be treated? The physician might treat the interdeterminate result in the same way he or she might treat a positive result, and proceed to surgery. In that case:

Positive (including indeterminate results)

Diagnosis	Prior	Conditional	Product	Revised
Cancer	3%	89%	267	10.7%
No cancer	97%	23%	2,231	89.3%
			$\Sigma = 2,498$	

The revised probability after a positive or indeterminate result would be only 10.7 percent, but 24.98 percent of patients would proceed to surgery (compared to 5.46 percent, as calculated above for positive FNA results) whereas 75.02 percent would avoid surgery.

In contrast, if an indeterminate result were treated as negative (i.e., the work-up stops), then the revised probability of cancer would be 0.4 percent because of false negatives. In this case, 82.78 percent of patients would avoid surgery (compared to 75.02 percent above).

Negative (including indeterminate results)

Diagnosis	Prior	Conditional	Product	Revised
Cancer	3%	11%	33	0.4%
No cancer	97%	85%	8,245	99.6%
			$\Sigma = 8,278$	

In a similar vein, consider the effect of an unsatisfactory FNA. Because cancers are more likely to provide satisfactory samples, an unsatisfactory result lowers the probability of cancer, as shown above, from 3 percent to 1 percent. Recall that a negative FNA only lowers the probability of cancer to 0.3 percent (because of the possibility of false-negative results).

5

Communicating with the Public about Exposure to Iodine-131

The U.S. Department of Health and Human Services (DHHS) asked the Institute of Medicine and the National Research Council to provide information that would help the department educate and inform the public and health professionals about the health implications of exposure to I-131 from the nuclear-weapons testing program in Nevada in the 1950s and 1960s. Chapter 4 discussed approaches to patient and clinician education and included sample information statements that might serve as a starting point for DHHS work to develop, test, and disseminate this kind of information. This chapter provides a broader framework for DHHS to consider as it develops its information strategy.

Communicating effectively with the general public about the health risks posed by exposure to I-131 fallout from aboveground nuclear tests some 40 years ago presents a difficult challenge to DHHS for several reasons:

• The aboveground nuclear tests in Nevada were purposive, man-made phenomena that left behind toxic residue. Since the tests ended, governments and residents of areas adjacent to the test sites have engaged in intermittent, often acrimonious, debate about possible health effects and about the release of information about the tests. The legacy is a government with a record of poor credibility as an information provider, and a subset of the population convinced that the health consequences of the tests are significant and severe.

• The best scientific estimates of exposure to radioactive iodine and of developing radiation-related thyroid cancer or other thyroid problems are burdened by significant uncertainties. These uncertainties must be explicitly considered in any effort to estimate the likelihood that a specific individual will be diagnosed with thyroid cancer or that a particular diagnosed cancer actually stems from radioactive fallout.

• Although a diagnosis of thyroid cancer must be treated seriously, it is an uncommon cancer and is rarely life threatening. Most individuals exposed to I-131 fallout from the Nevada weapons tests—even those few whose doses approached or exceeded 1 gray (Gy, 100 rad) are unlikely to develop exposure-related thyroid problems. Communicating small probabilities and changes in small probabilities may be difficult.

• A widespread attempt to alert the public to the possible health consequences of exposure to I-131 fallout and to promote systematic screening for thyroid cancer might not lead to entirely benign outcomes. No evidence shows that screening for thyroid cancer is effective in improving survival for this highly survivable disease. Screening tests are imperfect and can cause harm in the form of unnecessary surgeries and other procedures, anxiety, insurability problems, and other problems. The quality of the tests also depends on the skill of the screeners, which can vary widely.

Despite the complexity of the topic, the uncertainty of estimates of exposure and of probabilities of developing cancer, and public questions about government credibility, DHHS must devise ways to communicate accurately, credibly, and effectively about its 1997 (NCI 1997a) report (taking into account sound criticisms of its methods, conclusions, and presentation). Media and other attention to the report and to the fallout issue more generally will undoubtedly attract interest and concern and lead some people to want to learn more about their own potential chance of developing health problems. The committee believes that DHHS must accept responsibility for helping people understand the possible relevance of the NCI report to their own circumstances but recognizes that the limits of available data and methods will make this difficult.

The following sections of this chapter discuss characteristics and principles of risk communication that emphasize how people construct their own judgments of risk; the importance of source credibility in those judgments; the probability that the audience for risk information will be heterogeneous rather than homogeneous, thus creating the need for a variety of different information efforts; and the need to promote involvement rather than exclusion of the public in the risk communication process. The chapter also explores some specific communication strategies aimed at both groups and individuals. The resources and effort expended on risk communication should be proportional to the potential for harm and the likelihood that the risk communication will be successful.

CHARACTERISTICS OF RISK COMMUNICATION

As is the case with most risks, communicating effectively possible health hazards stemming from exposure to I-131 fallout from the Nevada nuclear-weapons tests will be complicated. Most risk communication efforts fail because communicators believe the process is relatively simple: "educate the public" (Liu and Smith 1990; Rogers and Storey 1987; Hyman and Sheatsley 1974). Informing the public is a useful goal, but there are contingencies that, if not accommodated, can frustrate even the most earnest effort.

The success of any message will depend on considering a variety of factors in designing an information campaign. Several of the important contingent conditions are listed here, both to illustrate what is available in the literature and to establish an interpretive framework that can inform DHHS public-education efforts. Success may be very generally defined as communication that is viewed as credible by intended audiences, that allows these audiences to accurately interpret factual information (e.g., probability statements), and that does not provoke unintended reactions (e.g., serenity when concern is warranted or vice versa).

What the Audience Brings to the Message

One mistake made by risk communicators is to assume that the audience is an empty vessel waiting to be filled with the communicator's interpretations of reality. The assumption of audience passivity is widespread, and it has found expression in such theoretical positions as the following:

• The powerful media hypothesis, sometimes called the hypodermic model, predicts that people generally are extremely vulnerable to media influence. It is the basis of most propaganda efforts and served as a catalyst for much of the communication research during World War II. Today it is resurrected whenever a new communication channel opens. The cyclical phenomenon is seen by looking back in time at public and policy reactions to the movies, comic books, radio, television, and now the Internet (Lowery and DeFleur 1995; Reeves and Hawkins 1986; Wartella and Reeves 1985).

• The third-person effect hypothesis states that when individuals encounter information that is potentially alarming, they suspect the information will strongly influence everyone but themselves. We believe ourselves capable of evaluating messages, but we do not ascribe the same skill to our neighbors (Perloff 1993; Davison 1983).

Audiences are far from passive and failure to recognize this can sabotage effective communication. People are more likely to bring interpretations *to* a message than they are to glean interpretations *from* a message (Zaller 1992; Derwin 1981). For example, Stevenson and Greene (1980) studied individuals' judgments about possible bias in news stories about presidential candidates and found that "people see as biased news information which is discrepant with the cognitions they already hold about the situation described in the news story" (p. 119). Put another way, a judgment of bias told the researchers more about the views of the individual doing the judging than it did about the content of a news story.

Communicators must learn that the audience participates in the construction of meaning. People bring complex sets of cognitions to any issue, including risks. If they attend to the new information, they will likely incorporate it in some way into their understanding of those issues. But that effort can change the meaning of the message more often than it will change the understanding of the issue at hand.

Application

The idea that audiences will be active in constructing understanding of an issue presents a challenge to information campaigners who wish to influence attitudes and behavior. With respect to I-131 exposure, DHHS must remain sensitive to this audience role and, like all information providers, it must make strategic decisions about how much of an impact it wishes to have and with whom. There are ways to work productively with active groups; some of them are discussed below.

One task for DHHS is to learn more about how members of the general public and more interested groups and advocates think about exposure to I-131 and the risk of developing thyroid cancer and to identify how they filter information through different experiences and conceptual frameworks. Ongoing evaluation of its communication efforts should allow DHHS both to learn more about public thinking and to assess how well its communication program is doing.

As noted above, the resources and effort expended on risk communication should be proportional to the concern about the risk as well as to the likelihood that the risk communication will be successful. It might be appropriate in some instances, therefore, to use less resource-intensive risk communication strategies. Whatever strategy is pursued, it must be remembered that people who encounter DHHS information about exposure to I-131 will predictably incorporate the messages in ways that are consistent with their personal understanding of the Nevada tests and their fallout. If the DHHS message is inconsistent with that understanding, the message will suffer.

Audience "Meanings" Will Vary in Theme and Intensity

There is no general audience for risk communication. Rather, any group of individuals will probably display a wide, often contradictory, array of meanings regarding a particular issue. A host of variables account for this variance including that people will differ in their knowledge about the risk at hand, in their motivation to learn more, in their capacity to learn more, and in the complexity of the belief systems they have constructed to make sense of threats generally.

Level of Knowledge

When there is a broad audience, it is difficult for communicators to judge what individuals already know about a particular hazard and its potential consequences. Studies of risk perception find a strong correlation between the salience of a risk and the amount of knowledge. During the 1980s, for instance, AIDS often topped the list of important problems for the American people, and respondents in surveys generally demonstrated reasonably high levels of knowledge about the risk (Fisher and Fisher 1992; Becker and Joseph 1988). Similarly, the summer drought of 1988 put global warming on the personal agendas of many individuals, likely leading to increased knowledge about that phenomenon and its implications (Trumbo 1995). It would seem, then, that for any risk, some number of individuals will have learned a great deal. Those persons also will tend to have established strong beliefs about that risk.

Motivations to Learn

Simply knowing little about a risk does not ensure that someone will work hard to learn more. People vary dramatically in their motivation to learn, but one strong incentive to learn is personal experience with a problem. Knowing someone with AIDS, for example, has long been a good predictor of information-seeking about that disease (Becker and Joseph 1988). The Centers for Disease Control and Prevention's Cancer Information Service offers another example. It experiences an upsurge in the number of telephone calls soon after news about a particular cancer is disseminated in the mass media (Freimuth 1998).

Sometimes, however, even highly salient conditions will not produce an effort to learn. We all make choices about what we do, and one option is to do nothing. For example, Wynne (1991) was at first surprised to find that apprentice workers at the Sellafield nuclear power plant in Great Britain knew little about basic radioactive processes and felt little need to know more. He and his colleagues subsequently learned that those workers had put their cognitive energies into learning organizational procedures, not science; instead, they placed their trust in the institution to protect them from harm. Dependency on their jobs may also have encouraged disregard.

Structural Limits to Learning

At times, even though motivation exists, an individual cannot summon the resources to learn. Socioeconomic status confers on some individuals an information-rich environment and on others an information-poor one. Sociologists (Tichenor and others 1980) hypothesize that structural differences can create a knowledge gap even in a society brimming with information. Given a societal need to know and the routine availability of information, they argue that people with greater resources will become increasingly knowledgeable while those with few or no resources will show little appreciable change for the better. Over time, a knowledge gap forms and widens.

Wynne (1991) reinforces this notion of structural limits to learning in his research on public understanding of a variety of risky situations in Great Britain. Specifically, "an important discovery from our research has been the enormous amount of sheer effort needed for members of the public to monitor sources of scientific information, judge between them, keep up with shifting scientific understandings, distinguish consensus from isolated scientific opinion, and decide how expert knowledge needs qualifying for use in their particular situation" (Wynne 1991, p. 117).

Complexity of Cognitive Systems

Confronted with a possible risk, people vary widely in the sophistication of their cognitive processes for making sense of risk. As in many other situations, people tend to employ judgmental shortcuts and personal theories of how the world works to decide what to make of a risk (Freudenberg 1992; Gilovich 1991; Kahneman and Tversky 1979). Those theories can be sophisticated or they can be naive. For example, some people have notions of causality that make chance an acceptable explanation for a cancer cluster; for others, assumptions that all effects have causes will preclude chance as a factor. As noted in Chapter 4, even basic literacy and numeracy must be recognized as challenges.

Application

Creating appropriate risk messages requires recognition of the heterogeneous nature of the audience and of the importance of risk salience, information resources, and other factors that inform people's responses to information. Information providers can profitably use survey research, focus groups, and strategic informants (including citizen advisory groups made up of representatives of a variety of concerned groups) to shape their understanding of these variables. Although the participation of concerned groups is important, representatives of the general public should also be included to ensure that the larger community feels neither ambiguous nor hostile to the communications and that its perceptions of risks related to I-131 exposure are understood.

Because people may vary in language and numerical skills or comfort levels, DHHS may want to develop materials that vary in scope and depth or that unfold in ways that allow people access to information at the level that is comfortable for them. On the Internet, for example, documents that are relatively short and simple can include links to more detailed information. Unfortunately, as discussed below, much remains to be learned about the Internet as a medium of communication. Also, those who might benefit most from this feature of the Internet may be among the income and educational groups that have the least access to the Internet. This is one reason why the Internet should not be the major communication tool.

Reinforcing Beliefs versus Changing Them

Most message designers try to change beliefs and behavior. The strongest effect of a message, however, can be to reinforce the status quo.

A typical example occurred some years ago when the government of Taiwan engaged in a national risk communication campaign to persuade residents that the construction of its fourth nuclear power plant was both safe and in the public's best interest. Underlying the campaign was the assumption that, as people learned more about nuclear power and the country's energy plans, they would develop beliefs in keeping with the campaign's goals. The idea backfired: Although many people believed they had learned more about the risks of nuclear power, they used that knowledge to strengthen their existing views. Residents who supported nuclear power before the campaign tended by its end to be equally or more enthusiastic about the technology; those who were skeptical about nuclear power at the start ended more strongly against the proposal (Liu and Smith 1990).

Of course, this reinforcing pattern works only when a person brings a belief system to the issue. When a risk is not well known or understood, information campaigns can have a substantial influence on the construction of knowledge and belief systems, and sometimes that effect will be consonant with the goals of the information provider. Even information touting the benefits of a technology, substance, or process over its risks can produce avoidance or hostility (Lopes 1987; Kahneman and Tversky 1979). Sociologist Allan Mazur (Mazur 1981) has examined this process and finds, for example, that "balanced" media accounts—which discuss the benefits and the harms of a process or technology—produce predominately negative reactions among people who are unfamiliar with the process or technology in question. And the more coverage provided by the media, the more negative public attitudes become (Mazur 1981).

Application

Traditional informational messages about the health effects of aboveground testing will serve to reinforce existing beliefs, whatever they are, among individuals who already have complex and enduring beliefs about those risks. Effective communication with these groups may require more intensive work such as community meetings, establishment of advisory groups, and working with other information sources that may be more credible to those with strong existing beliefs. Other people will perceive they know little and, if they are motivated to learn, will be much more open to DHHS's attempts to frame the issue for them. News releases, Internet communications, and other more traditional approaches may work well with this audience.

Risk Perception

One finding from the literature is that individual risk judgments are formed by many factors. This contradicts the common supposition that an informed risk judgment is one that achieves a tight fit between the likelihood of coming to harm and individual willingness to take preventive action. As many researchers have demonstrated, the possibility of harm does contribute judgment, but its relative weight depends on a host of other factors, among them assessments of a risk's controllability, its magnitude, the extent to which a hazard might harm future generations, whether or not it is voluntarily assumed, and its ability to harm non-human organisms (Slovic 1992; Slovic 1987; Johnson and Tversky 1984).

Some risks seem inherently unnerving; others do not. This fact has led scholars to argue on behalf of a cultural basis for perceptual differences. That is, a particular interpretive framework can be nurtured among groups of people such that the framework defines some risks as more terrifying and less palatable than others regardless of the risks' other attributes (Douglas and Wildavsky 1982). Kasperson and colleagues (1992) articulate a process, called the social amplification of risk, to explain this. In their view, "events pertaining to hazards interact with psychological, social, institutional, and cultural processes in ways that can heighten or attenuate perceptions of risk and shape risk behavior" (pp. 157-158).

Public perceptions of nuclear power likely have been shaped by this complex amplification process. For example Slovic and colleagues find the term "nuclear" is a strongly aversive term for Americans, evoking feelings of peril and ugliness (Slovic and others 1991). Weart (1988) contends that these nuclear fears are rooted in our social and cultural consciousness, and Peters and Slovic (1991) offer evidence that attitudes toward the world and its social organization (called "world views" by scholars who study the phenomenon) serve as "orienting dispositions" that guide individuals' responses to all things nuclear. Regardless of cause, Americans' affective response to nuclear technology is so intensely negative, say these scholars, that it overwhelms any positive affective response. The "nuclear connection" may have a strong negative impact on individuals' thinking about the health risks of I-131 independent of health issues themselves.

Application

Deciding what to emphasize in messages about risk depends on intelligent assessments of what dimensions of a risk matter to the audience. The aboveground nuclear-weapons tests will provoke cultural reactions as well as disparate individual responses. DHHS may need to focus on more than I-131 exposure and probabilities of developing thyroid cancer and to acknowledge that radiation is not like most risks because it is not observable and exposure is not voluntary or often even known by those exposed.

Information Channels

As individuals, when we think about risks, we make important distinctions between our personal risk of coming to harm and the likelihood that other people will be harmed. Specifically, we view ourselves as less vulnerable; everyone else is at greater risk than are we.

The distinction between "me" and "them" pervades life—not just risk estimates. So it makes sense to explore the possibility that we all use information differently to inform our understanding of "me" and "them." Such a distinction is particularly important to the success of information campaigns, where the goal is typically influencing individual perceptions of personal situation, often to the extent of inducing a change in behavior.

Numerous studies find that we seem to differentiate between channels of information that we find appropriate to informing our understanding of "them" and those acceptable for an understanding of "me." Specifically, we value mediated information channels—newspapers, television, magazines, radio—as sources of information about other people while we resist the relevance of the information channel to our own understanding of our situations. Indeed, we rely heavily on interpersonal channels—people we trust—when we need to make personal choices.

This tendency is called the impersonal impact hypothesis (Mutz 1992; Tyler and Cook 1984) and it seems to be pervasive. Illustrative of findings in this arena are those that examine influences on individual perceptions of victimization by crime. Such studies typically find that people use media accounts of crime to inform their understanding of the prevalence and threat of crime in society generally but that they do not interpret those stories as telling them anything about their likelihood of personally being victimized (Tyler 1984; Tyler 1980).

Similarly, we prefer to use media accounts of risk to inform our general, societal understanding of a risk, but rarely do we employ these accounts to inform our personal risk situations. Put another way, the primary effect of media accounts of risk seems to be knowledge gain (see, for example, Schooler and others 1998). But when it comes to making a judgment about personal risk, media accounts are insufficient sources of information. Instead, we need to talk to someone (Dunwoody and Neuwirth 1991).

Application

It is important to distinguish between individual personal risk judgments and perceptions of the risk to others; some channels of information will be more appropriate for one than for the other. If DHHS is trying to communicate information about the general risk stemming from the Nevada tests, mediated channels are fine. There is some evidence that individuals who attend to those messages will learn from them. But once people begin to seek information to inform their personal situations, other channels, interpersonal ones, must come into play.

Source Credibility

People make the acceptability of information a function of the perceived trustworthiness of the source. Simply put, untrustworthy sources will not be believed, regardless of the quality of evidence they present about an issue.

Freudenburg (1992;1988) attributes this heavy reliance on trustworthiness as a heuristic device to "recreancy" and sees it as a normal by-product of an increasingly complex society. Formally defined, a recreant is someone who fails to do his or her duty or to be faithful to his or her obligations. We have become heavily dependent on specialists to make society run, argues Freudenburg. Because the typical citizen cannot evaluate the detailed work of specialists, he or she must resort to making judgments about the ability of the specialists to do their duties carefully (Freudenberg 1992; 1988). And all it takes to generate a diagnosis of untrustworthiness within a complex system, says Freudenburg, is one mistake, one display of carelessness or ignorance. Because the typical American cannot know a system intimately enough to interpret errors as more or less serious, he or she instead defines any error as proof of recreancy, and credibility suffers. Freudenburg argues that a public perception of recreancy explains more of the variability in public concern with such hazards as nuclear power and nuclear waste than do such characteristics as sex and socioeconomic status.

Other scholars, among them Kasperson and others (1992) and Peters and others (1997), offer sets of underlying dimensions for trust judgments that include perceptions of a source's competence, objectivity, fairness, and compassion.

Typically, we assume that trustworthiness judgments are directed toward the individuals or organizations who are sources of information. That might not always be so. People sometimes rely on judgments of the credibility of information channels rather than of information sources. The choice can depend on the complexity of the belief system that a person brings to the message.

The Person Who Knows Little about the Risk

If someone knows little about a risk it suggests that the risk is not highly salient (at least at the moment). When someone has less at stake from an information perspective then he or she will invest less effort in making a credibility judgment about available information. One way to invest less effort is to employ judgmental shortcuts that focus on the credibility and trustworthiness of information channels, not information sources. That is, rather than enter into the labor-intensive task of picking apart a message to judge the credibility of a particular source, an individual will make a broader judgment about the credibility of the channel. In this context, a source is the originator of information and a channel conveys the information: A newspaper story (a channel) carries information from a government official (a source).

Like all heuristic devices, this default to channel is efficient. Most of our information about risks comes to us through some mediated channel. And a channel judgment often is grounded in culturally negotiated notions of channel quality; for example, most people will regard NBC television as a more credible channel than the *National Enquirer.*

The Person Who Has Developed a Complex Belief System about the Risk

People who have enduring belief systems will already have defined the risk in question as highly salient, will probably be willing to engage in more systematic information processing when they see messages about it, and thus will make much finer credibility distinctions among the sources than solely among the channels. They will not simply judge a risk on the basis of the channel itself but will instead attempt to learn who is speaking with authority and to decide whether the authority is warranted.

Information sources bring their own cultural baggage along. Some recent research indicates that general cultural expectations of sources influence our perceptions of credibility. For example, Peters and others (1997) found that the mix of underlying predictors of trust and credibility varied by type of source: The two strongest predictors of the credibility of government sources were the perceived commitment to a goal (such as public health) and knowledge and expertise regarding the risk. The two strongest predictors of trust in industry sources were the perceptions that the industry was concerned about the risk and that the industry sources tried to provide full information quickly.

Evaluating the role of trust and credibility is made even more complex because of the possibility that credibility judgments are at least partially contingent on the source's area of expertise. In a survey of perceptions about a variety of risks among residents of 5 European countries, Jungermann and others (1996) found that those surveyed clearly regarded sources that are critical of industry (environmental groups, for example) as the most trustworthy. But when they were asked to match specific types of risk information with the ideal source for that information, respondents indicated that even low-credibility sources could provide some trustworthy information. For example, although they were critical of industry sources, survey respondents regarded industry as the best source of information about the characteristics of their own products. Government sources, another category with credibility problems in this survey, were regarded as the best sources of information about relevant environmental regulations.

Application

If a source or channel is not trusted, it is almost useless as an information provider. And in such long-running and volatile issues as the aboveground Nevada tests, credibility is important in public acceptance of information. In lieu of changing perceptions of credibility—a decades-long task at best—it might be useful to separate types of information such that a specific type of information is matched with an information provider who will be seen as credible in that instance (see Jungermann and others 1996). For example, a federal agency, such as NCI, might be defined by the public as a highly credible source of information about how to calculate the risk of developing thyroid cancer from exposure to I-131. But NCI's expertise might not give it legitimacy as a source of information about how concerned any individual should be about that risk. By involving members of the public in developing and testing information, DHHS may also identify other, more credible information sources (e.g., some state health departments) and encourage their involvement in providing information.

Communication as a Two-Way Process

Risk managers now understand that communicating successfully with a set of constituents means involving those constituents in the process much earlier than was done previously. Risk managers of old viewed communication as a one-way, temporal educational process in which experts pronounced and the audience listened and learned. An unsuccessful risk communication effort was blamed on audience failure to assimilate the information and act appropriately.

That unilateral approach is slowly being replaced by a multilateral one as agencies with a communication mission work to involve the audience in planning and execution. Serving as an important catalyst for this change was the National Research Council report *Improving Risk Communication* (National Research Council 1989), which declared "risk communication should be a two-way street," between experts and various groups, that should exhibit a spirit of open exchange in a common undertaking rather than a series of "canned" briefings restricted to technical "nonemotional" issues and an "early and sustained interchange that includes the media and other message intermediaries" (p. 10).

In response, many state and federal agencies are beginning to assemble citizen advisory groups and to meet and seek the counsel of the public in the course of designing risk messages (see, for example, Boiko and others 1996). There are good reasons for this. First, individuals who have developed complex belief systems about a risk situation tend to be highly knowledgeable about the risk and often have developed insight about the risk perceptions of various interested groups. They can thus bring useful information to the table. Second, involving citizens early in a risk communication process enhances interpersonal communication of the type most likely to influence the perceptions of even the most hostile individual. Numerous efforts are under way to engage citizens in interpersonal dialogue about issues to enhance decisionmaking. The most visible of these recent efforts was the 1996 National Issues Convention. Before the 1996 presidential election, a random sample of voting-age adults assembled in Austin, Texas, to discuss issues with policy makers and with one another. The object was to encourage people to deliberate and negotiate with others, and subsequent surveys of the participants indicated not only that they learned a great deal about the issues under discussion, but also that many of them did indeed change their understanding of the issues. An added benefit was an enhanced sense of political efficacy; these persons now believed their opinions mattered (Fishkin 1996).

As judged by the degree of community involvement, perhaps one of the better current risk communication templates is the Hanford Health Information Network, sponsored by the state health departments of Washington, Oregon, and Idaho. Like the Nevada tests, the federal Hanford Nuclear Reservation is the subject of lengthy and continuing public concern about the health implications of radioactive contamination. However, the Hanford situation is sufficiently different from the Nevada Test Site (e.g., range of hazardous materials and exposure routes involved, time period, geographic area involved), that the various elements of this model and the kinds of information provided would need careful examination.

Application

If DHHS deems this particular risk worthy of a sustained campaign to inform not just the general public but also various interested groups, it will be important for the department to design information structures that bring the public into the process early. The "stakeholders" are the people who have developed complex belief systems about the risks posed by the Nevada tests or similar events and those who are most at risk, regardless of their a priori knowledge and beliefs.

CAN RISK BE COMMUNICATED EFFECTIVELY?

A simple, albeit ambiguous, answer to this question is "Yes." The difficult part is making educated guesses about how to accomplish the goal. There has been some empirical exploration of how to communicate about risk with nonspecialists—studies of risk comparisons, of explanatory devices, of appropriate information channels—but the work is relatively sparse and one must generalize with care. Equally challenging for risk communicators is the task of selecting the interpretive context within which facts will be presented. That is, should one talk about an individual's immediate or long-term prospects? Should one talk about the risk to the average person or instead emphasize contingent conditions that would make some people much more susceptible than others?

The remainder of this section briefly discusses elements of the interpretive context, offering a list of domains that DHHS should consider as it decides on the content of its messages. The use of specific communication strategies is discussed, including risk comparisons, the question of whether the World Wide Web can contribute, and a means for people to estimate personal risk from exposure to I-131 during the Nevada tests. Finally, a recommendation is made for how to accomplish communication planning by bringing the affected public into the process.

Appropriate Comparisons

In the course of studying the health risks created by the Nevada atomic weapons tests, this committee has come to several conclusions that need to be reflected in the communication strategy developed by DHHS.

• Being at increased risk of developing thyroid cancer has little influence on individual mortality. This is true for two reasons: Thyroid cancer is rare, and long-term survival after diagnosis is high. No evidence suggests screening would improve these already high rates. It will be important to account for contingent conditions when trying to evaluate an individual's future lifetime risk of being diagnosed with thyroid cancer. Such evaluations should emphasize that increased probability of developing thyroid cancer caused by I-131 exposure is restricted to a relatively narrow birth cohort and that, even for members of that cohort, future risk of developing thyroid cancer is most likely less than a few percent. For the public, explanations of probabilities of developing or dying of thyroid cancer related to I-131 exposure should be provided in both qualitative and quantitative terms as described in Chapter 4 and explanations should include baseline risk (e.g., risk of naturally occurring thyroid cancer).

• Risk estimates will come bearing high levels of uncertainty that will complicate efforts to explain probabilities of developing thyroid cancer related to I-131 exposure. Although NCI (1997a) has estimated as much as a two-thirds increase in risk for developing thyroid cancer for selected birth cohorts (people born between 1948 and 1952, for example), this estimate of increased risk is accompanied by a wide range of uncertainty. Moreover, available epidemiologic evidence so far provides scant support for the higher estimates of excess cancer cases.

As suggested in Chapter 3, if individual exposure to I-131 from fallout can be accurately assessed, then for a diagnosed case of thyroid cancer, the probability that the cancer was caused by fallout can be assessed.

SOME IMPORTANT COMMUNICATION ISSUES

Risk Comparisons Are Risky

One obvious way to place a risk in context is to compare it with a second, third, or fourth cause of harm. For example, Chapter 3 compares thyroid cancer incidence and mortality with comparable statistics for other cancers.

When scientists and risk managers attempt to put risks in context by making comparisons, their audiences may derive meanings from them that are quite different from what was intended. For example, when one scientist equated the likelihood of coming to harm from eating contaminated fish from the Great Lakes to the risk of getting cancer from exposure to chemicals found in the food, drinking-water, and air of some of America's more polluted cities, his attempts were interpreted as venal efforts to downplay the risks of eating fish (Dunwoody and Peters 1992).

Although comparison is an excellent explanatory technique, it works only if it allows the user to connect appropriate dimensions, such as the quality of life, associated with different health outcomes. A scientist will offer a risk comparison with the intention that the user will compare the likelihood of one cause or another leading to harm. But a lay user of a comparison might attempt to make comparisons between other dimensions of the risk, for example, whether the risk is voluntary or involuntary. Most people consider multiple dimensions when making risk judgments, and therein lies the problem. In the example above, the scientist intended the comparison to illustrate the magnitude of various risks that are commonly encountered and accepted. But the audience inferred that the scientist was comparing dissimilar values. If a source's credibility is under attack, risk comparisons are problematic. When a source that is deemed untrustworthy offers comparisons, those comparisons can be interpreted as attempts to persuade rather than to inform.

Freudenburg and Rursch (1994), who have conducted one of the few empirical explorations of the value of making risk comparisons, add that the use of comparison can be successful explanatory strategy only when the comparison is by a source that is trusted by the audience. The authors note, "Essentially, all risk comparisons—even those that are widely seen as acceptable risk comparisons—involve the provision of information by proponents and officials. If those proponents and officials are not trusted, then even 'legitimate' risk comparisons can do more to arouse suspicions than to assuage them" (p. 954).

Application

A comparison of estimates of harm works only if the audience interprets the comparison as intended by the communicator. One way to achieve that is to identify risks for comparison that are similar in areas that are important to risk judgment, for example, the extent to which the risks are voluntary and the extent to which they are familiar. Such a strategy could, however, narrow comparisons substantially and make it difficult for people to put a risk in context. For radiation-related risks, identifying even this narrower set of appropriate comparisons is difficult.

Thus, DHHS may want to test how different kinds of comparisons involving radiation-related risk are perceived by different audiences. A later section of this chapter returns to this issue.

Explaining the Concept of Risk

Most of us try to explain complicated things by a largely intuitive process. We employ simple words and short sentences. We use active voice verbs, analogy, and metaphor. But sometimes this is not enough. Scholars in science education have become specialists in explaining science to the public; they help educators and communicators develop a keen focus on how audiences interpret information. For example, Rowan (1990) suggests that even the most carefully selected analogy or definition will fail if it collides with an incompatible belief held by the audience member. Beliefs inconsistent with theory are rife in daily life. If an individual happens to believe that a co-occurrence of apparently highly unlikely incidents—for example, the diagnosis of the same rare illness in three players on a football team—simply cannot occur by chance, then even the most carefully articulated explanation of the role of chance in such an event will not be credible. Rowan (1990) notes that an explainer must first refute the "naive theory," in this case by helping someone understand that the occurrence of a rare illness in three proximate individuals is not nearly as unlikely as originally supposed. Only then can the explainer safely offer the new theory.

In addition to naive theories, audience members can come to a risk message with only the barest numeracy skills, making it difficult for them to interpret quantitative representations of risk. In a study of adult women in New England, for example, Schwartz and others (1997) found that numeracy skills were strongly predictive of the ability of women to accurately interpret information about mammography and risk reduction.

Application

Explaining scientific concepts and processes and representing risk numerically requires an understanding of what the audience knows or believes to be true. If numeracy skills are low or if individuals have accepted competing naive theories the risk communicator must tailor the message accordingly, and consider the possibility and consequences of communication failure.

World Wide Web Communication

The World Wide Web's popularity has led many to predict an explosion in public reliance that will eventually exceed that for today's traditional information channels, with the possible exception of television (CommerceNet 1997; WorthlinWorldwide 1996). Indeed, although the Web is a fundamentally new kind of information channel that could be ideal for communicating some information, it is important to be cautious about using it as a primary channel of information for at least two reasons.

First, it is not yet a mainstream channel. Although use of the Web is spreading among the American population faster than did the telephone or television, it is still available mainly to people who have regular access. This excludes a large portion of the public. The emergence of technologies such as the Web TV will increase the rate of use, but for the near future access will be available only to a minority of the U.S. public (Hoffman and Novak 1998). An information manager who is responsible for communicating with a wide audience is not well served by employing only the Web alone.

Second, we are just beginning to explore the role of the Web as a public-information channel. Research on hypermedia sites exists but has looked primarily at use of such channels for formal education purposes. We have no idea how the Web will be used as an informal channel, but it would be fair to expect that such uses would be very different from those in the classroom. Is the Web a mediated channel or an interpersonal one? If it emphasizes nonlinear communication—by urging the user to dart from one place to the next—what is the nature of the learning that takes place? How do users make reasonable judgments about the trustworthiness of the information presented?

Although research on how people use the Web as an information channel is just beginning, one possibility, admittedly optimistic, is that, for risk communication purposes, the Web combines the best attributes of mediated and interpersonal channels. As a mediated channel, it can reach millions of people inexpensively. And its interactive qualities can encourage users to interpret a site's contents as telling them something about themselves personally—a conclusion usually reserved for interpersonal channels.

If users are willing to extrapolate site content to their personal situations— something most people are extremely unwilling to do with information encountered in the mass media—then the Web could become an important means through which to teach users about a risk and help them make personal risk judgments.

The National Cancer Institute seems to be actively considering this possibility. Its own Web site already contains a section that allows a user, given the relevant background and behavioral information, to calculate his or her own likely dose of I-131 from Nevada Test Site fallout. The section is not very accessible, however, and it could frustrate users more than it helps them.

Application

DHHS should actively and creatively use its Web site but not regard a Web site as its primary communication tool. The Web should supplement other information channels, but it cannot replace them. The committee has found no empirical evidence to support the idea that calculating one's personal risk from information and a formula available on the NCI site will be helpful to members of the general public. But a combination of careful information choices and thoughtful design might make such a Web site worth trying and evaluating.

PUBLIC INVOLVEMENT IN EXPLAINING RISK

Many scholars, and an increasing number of risk managers, encourage a differentiation between technical and democratic approaches to risk communication. As described by Rowan (1994), the technical perspective is characterized by "the use of physical and social sciences to describe and predict health effects posed by environmental hazards" (p. 392). Such risk assessments construct a picture of some hazard's health effects and are used to set safety standards. People who subscribe to the technical approach, says Rowan, put their faith in scientific expertise. Those who subscribe to the democratic approach, on the other hand, are concerned about the political fairness of the assessment of risk. Notes Rowan (1994), "The democratic view is that who decides whether some risk should be incurred and who benefits or is harmed are just as important, if not more, than the severity and likelihood of the risk" (p. 396).

Communication strategies are different for the two approaches. To facilitate a technical view, one would use an expert-to-novice flow of information. The democratic approach requires that all parties affected by a risk are guaranteed participation and power in decisionmaking. Each strategy has limitations. For example, the democratic approach cannot be implemented retrospectively, and thus, in the current instance, this strategy will be heavily burdened by past decisions relative to weapons testing and the resulting exposure of the public to I-131 and to other fallout radionuclides. The democratic approach now can be used only to guarantee that all affected parties can participate and affect future decisionmaking.

The issue of health effects caused by I-131 fallout from the Nevada tests is one for which technical and democratic perspectives are both present and, at times, seem to collide. DHHS can choose to communicate with the public only with the technical approach. But the department must realize that such a choice will be quite unsatisfactory for some subgroups of the public. Communicating effectively with these often highly charged subgroups may well be expensive and time-consuming. If DHHS wishes to promote a public perception—particularly among these subgroups—of the government as trustworthy, it will be a challenging undertaking. To that end, this committee provides a set of recommendations for how to achieve a communications arena in which all parties may participate.

Application

Effective communication by a government agency about public exposure to radiation from federal facilities must account for the public perception of radiation as a health threat and the history of less-than-forthright communication about such exposures. To establish its credibility as a source of information, DHHS will need to engage interested groups—affected members of the public and the health professionals who serve them—as partners in the development and dissemination of information materials. Because interest in the health effects of I-131 exposure may continue for some time, attention should be paid to maintaining and improving—not just initially creating—the infrastructure to communicate information to the interested parties. DHHS should consider the following process of developing and communicating information about the health effects of exposure to fallout from nuclear weapons testing at the Nevada Test Site. DHHS will also need to consider the effectiveness and cost of different approaches as it develops its strategy.

• Development of Public Information on Radiation Health Effects.

To promote the provision of consistent public information about exposure to I-131 from the Nevada weapons tests, DHHS should consider creating a national resource center to help inform people concerned about their individual risk from such exposure. Ideally, the center would be under the guidance of an organizational unit familiar with radiation health effects and with communicating such information to the public, for example, the Centers for Disease Control and Prevention. If such a resource center were created, DHHS should also consider establishing a council to provide guidance on the center's communication methods and materials. The council should consist of individuals selected to represent the concerned public, including the "downwinder" organizations and Native Americans groups; physicians; public-health authorities; and scientists knowledgeable about dose reconstruction, risk assessment, epidemiology, and risk communication. Before they are published, materials should be reviewed by a panel of technical and lay advisors.

Public communication materials should be developed to include at least the following items:

—a description of the NCI (1997a) report and its limitations;

—a description of the report's focus on I-131 rather than on other isotopes;

—a description and explanation of the factors that place an individual at risk, e.g., age at exposure, sex, source and volume of milk consumed;

—a method to help people place their risk in perspective, again with adequate explanation of the uncertainties in individual estimates of risk;

—a description of the possible health effects and their likelihood of occurrence;

—a description of the limitations and risks of harm associated with screening for the health effects of radiation exposure; and

—recommendations about factors people should consider in deciding whether to seek further information or advice.

The addendum to this chapter includes for DHHS consideration an example of a method people could use to assess their exposure to iodine-131 and their risk of developing thyroid cancer and to help them decide whether they wanted to seek further information and advice. The specific county classifications, risk categories, and multipliers provided for in the assessment method would need to be chosen carefully and validated appropriately. The method should provide qualitative as well as quantitative information and should include information on baseline as well as relative risk. The inherent limitations of individual dose estimates should be clearly explained. The general approach illustrated in the addendum would replace the unwieldly method for individual dose assessment currently included on the NCI's Web site for the I-131 report.

- Development of Information for Health Care Providers.

Of equal importance to public information is the development of information for health care professionals. Chapter 4 discussed the kinds of information needed by physicians and presented a sample information sheet that DHHS might use as a starting point. Consistent with the suggestions above regarding public communication strategies, it recommended that professional groups and health care organizations be involved in developing, testing, and disseminating information materials through a variety of channels including professional meetings, journals, education videos, and public and private Internet sites. The information for clinicians described in Chapter 4 builds on the chapter's assessment of the evidence about the benefits and harms of thyroid cancer screening. It recommends against a program of routine screening. If DHHS chooses to disregard this evidence-based assessment and substitute a political judgment, materials provided to clinicians and others should make clear that there is a lack of a scientific justification for routine screening of the general population or potentially exposed population subgroups.

- Distribution of Information.

In developing a method that could be used to distribute information, one approach it might consider is funding the assignment of public-health personnel educated on the topic of Nevada Test Site radiation health effects in some or all public-health-service regions. These personnel would provide information to interested members of the public, as well as to state health agencies, and would respond to questions and concerns. In general, plans for communication to the regional audiences should be developed at the regional level. These plans should be provided to the interested public and to state health agencies for input on their adequacy in meeting local needs. Special attention should be paid to identifying the most effective means of communicating with the interested parties, such as establishing a user-friendly Web site and a toll-free telephone information line staffed by skilled risk communicators who are knowledgeable about the health risks of exposure to radiation from the nuclear weapons tests. At the national level, DHHS should consider sending information to publishers of wellness newsletters and to organizations that represent affected groups.

CONCLUSIONS

This chapter has provided a broader framework for DHHS to consider as it develops its information strategy. It notes the challenges that DHHS will face in communicating complicated information about nuclear fallout, radiation exposure, probabilities of developing thyroid cancer, and clinical and public health responses. It also notes the challenge of communicating against a historical backdrop of controversy over government actions and information. Involving the public in the further development and testing of its information program is one way for DHHS both to better understand public concerns and to develop information that is credible and understandable.

ADDENDUM 5: EXAMPLE OF A METHOD TO ASSIST INDIVIDUALS IN ESTIMATING THEIR PERSONAL THYROID CANCER RISK

The IOM/NRC committee believes that some Americans will want to learn more about the thyroid cancer risk they individually face. In this context, *risk* refers to the chance that a specific person will develop thyroid cancer as a result of exposure to I-131 from the Nevada tests. As discussed earlier, the lifetime risk of developing thyroid cancer is, even for most people exposed to I-131, still small, and the disease is rarely life-threatening. About 90 percent of people diagnosed with papillary thyroid cancer are alive 30 years after diagnosis. Any method to estimate I-131 exposure and the chance of developing thyroid cancer should be accompanied by clear information about these points and about the potential benefits and harms of screening for thyroid cancer.

Each person's exposure to I-131 and his or her chance of developing thyroid cancer are usually very difficult to determine as well as difficult to communicate. The method developed by NCI to help people assess their exposure is extremely complicated involving complex calculations for each of the test series and recollections about events that are decades past. It does not provide information about cancer risk.

The IOM committee believes that if DHHS wants to make available a method for assessing exposure and cancer risk, the method should be simple for the average person to use. The method presented here illustrates one approach that could be presented in a brochure or Web site along with the kind of information about thyroid cancer and thyroid cancer screening that is reviewed in this report. This method and related information would need to be tested with potential users.

Though the committee has reasoned that the individual county average dose estimates are imprecise to a degree that they are not useful to assess risk to individuals, there are more readily determined factors about each person that are related to their risk.[1] Individual thyroid cancer risk is determined by several factors beyond simply where a person lived during the time of the Nevada nuclear weapons tests. People exposed to I-131 can be found in almost any part of the United States. Gender, year of birth, some lifestyle characteristics, in particular, the rate of consumption of milk, and the source of their milk have to be considered too. Although age and gender are easy to determine, dietary recall informa-

---

[1]An alternate wording is to say, ". . . there are factors describing each person that are related to the likelihood they will develop cancer as a result of their exposure to I-131 in radioactive fallout."

tion—for example, recall of milk consumption some 40 years past—is generally viewed as unreliable.

The method described here is neither fully developed nor tested and this should be understood. Several steps by DHHS would be required as part of its development. Foremost would be to classify the 3,000+ counties into one of 3 groups that would differentiate between concentrations of I-131 in commercial milk available in the county (averaged over the testing period), and into one of 3 groups that would differentiate the counties on the basis of the concentration of I-131 in milk obtained from backyard cows (averaged over the testing period).[2] These two lists would be necessary companion tables to the method. In addition, supplementary material would also be necessary to provide information to the reader about the nuclear testing program, the years of birth that would result in the highest risk, thyroid cancer risks specific to each sex, radioactive fallout and I-131, natural occurrence of thyroid cancer, the benefits and possible harms of thyroid disease screening, the prognosis of thyroid cancer once it is diagnosed, and information about seeking counseling and guidance from a family physician or other medical specialist.

Again, the intent of the committee is to provide an example of a relatively simple procedure to assist in assessing and communicating an individual person's risk of developing thyroid cancer following exposure to the Nevada tests. In addition, the method should be accompanied by information about thyroid cancer and the potential benefits and harms of thyroid cancer screening similar to that described in Chapter 4. The committee expects that the brochure material, of which this would be a part, could be made available to those persons seeking information from DHHS about their risk related to the Nevada test. The committee does not suggest using the method as a mass screening device.

The following steps present the method as it might be written in a brochure.

HOW YOU CAN ESTIMATE YOUR RISK FROM EXPOSURE TO I-131 FROM RADIOACTIVE FALLOUT DURING THE 1950S

Step 1. Determine whether risk is likely.

Procedure: If you were born after 1910 but before 1960, you might be at some risk of fallout-related thyroid disease. Continue to Step 2. If you were born

---

[2]A note of explanation is required here in defense of the construction of this method. Though the committee feels the individual county dose estimates represent an over-statement of precision, the development of three categories would effectively collapse the wide range of individual county estimates (iodine concentration in milk or average dose) into three general categories: low, medium, high. The committee believes that this level of discrimination between counties is possible and defensible. We refer the interested reader to Beck and others (1990, p. 571) for a map of cesium deposition over the continental United States as an example of the level of discrimination among counties that is suggested here.

at another time, you do not need to continue further. Your risk from exposure to I-131 released by the nuclear weapons tests is near zero.

Explanation: Because age at exposure is the major determinant of risk, the first step is to decide whether there is an appreciable individual risk, based on age at the time of the nuclear testing program. For anyone born after the completion of testing, the risk as a result of the nuclear testing would have been zero,[3] that is, there would have been no exposure to the I-131 released from the tests. Equally, for anyone over the age of 20 years in 1951, although there might have been exposure, the risk would have been miniscule and would have been essentially zero for anyone over the age of 40. Thus, there is a potential risk for people born after 1910 and before 1960. Anyone else has zero risk, and need not continue with this procedure.

Step 2. Obtain multiplier factors from Table 1A or 1B.

Procedure: If you drank commercial milk routinely, obtain a factor from Table 1A. If you routinely drank milk from your or your neighbor's backyard cow or goat, obtain a factor from Table 1B.

Explanation: The NCI (1997a) report shows that the most important route of exposure to I-131 was milk consumption. The contamination of milk was a result of the contamination of the pastures where the animals fed. Because milk is not always consumed at or even near where it is produced, the distribution system for commercial milk was one factor that determined the variation in doses across the United States. If milk was supplied by a backyard cow or goat, the main determinant for dose was the amount of fallout where the animal was pastured, presumably near the home of those who consumed the milk. Thus, the second step in the self-assessment is to determine whether the main source of milk was a commercial source (A) or private (B).

Those who consumed mostly cows' milk from a commercial source during the testing refer to Table 1A, which is derived from the NCI data using as a reference group, males aged 5-14 who drank average quantities of milk during the 6 major test series between 1951 and 1957. Commercial milk supplies were not necessarily produced and consumed in the same counties, the geographic distribution of exposure reflects the pattern of distribution.

Milk from a backyard animal was likely to be more contaminated because of the much shorter time between milking and consumption and because of the possibly wider area over which such animals grazed due to the smaller amounts of supplemental (uncontaminated) feed provided. Exposure is related to deposition

---

[3]There is a baseline risk of thyroid cancer as discussed in Chapter 3. The *Percentage Lifetime Risk* in the absence of radiation exposure (over that of natural background radiation) is 0.42 percent for males and 1.07 percent for females. These concepts are described in Chapter 3 and the Addendum to Chapter 3 of this report but would require elaboration in the public materials.

TABLE 1A Multipliers for County of Residence at Testing for Those Drinking Commercial Milk[a]

COUNTY[b]	MULTIPLIER[c]
Low	CO_L
Medium	CO_M
High	CO_H

[a] If no milk was consumed, use a multipler of ___ (this value accounts for inhalation and other minor exposure pathways).
[b] List of counties to be supplied by DHHS.
[c] Abbreviations for variables are provided here; numerical values would need to be developed.

in the area. Thus, to determine relative risk according to location, use is made of Table 1B, in which the counties are divided into 3 categories—low, medium, and high deposition. The allocation of counties was determined for males, drinking average quantities of milk from a backyard cow summed over the 6 major test series between 1951 and 1957. The multiplier from Table 1A for the county of main residence during the testing is thus selected as the starting point for the risk self-assessment. Table 1B is analogous to Table 1A but the distribution of counties in the low-, medium-, and high-deposition categories is somewhat different.

Step 3. Obtain factors from Table 2.

Procedure: Obtain 3 factors from Table 2: A factor for age at the time of the tests, a factor for your gender, and a factor for milk consumption rate at the time of the tests.

Explanation: Several factors are important for assessing individual risk. Age is a major determinant both because the young thyroid is more sensitive to the effects of radiation and because the younger the child, the smaller the mass of the

TABLE 1B Multipliers[a] for County of Residence at Testing for Those Drinking Milk from Backyard Cows or Goats

COUNTY[a]	MULTIPLIER[b] for Cows' Milk	MULTIPLIER[b] for Goats' Milk
Low	CM_L	GM_L
Moderate	CM_M	GM_M
High	CM_H	GM_H

[a] List of counties to be supplied by DHHS.
[b] Abbreviations for variables are provided here; numerical values would need to be developed.

TABLE 2 Multipliers for Age at Testing, Gender and Milk Consumption

FACTOR	MULTIPLIER[a]
Age during Nuclear Testing (select your year of birth)	
1955-1960 (0-4 years during second half of test series)	A_{0-4}
1950-1955 (0-9 years)	A_{0-9}
1945-1950 (5-14 years)	A_{5-14}
1940-1945 (10-19 years)	A_{10-19}
1935-1940 (15-24 years)	A_{15-24}
1910-1935 (25-39 years)	A_{25-39}
Milk Consumption Rate (typical per day)	
None	CR_0
Average (0.5 to 0.7 quart/day [or liter/day])	CR_{avg}
Twice or more the average	CR_{2x}
Gender	
Male	G_M
Female	G_{FM}

[a]Abbreviations for variables are provided here; numerical values would need to be developed.

thyroid, which leads to a higher concentration in the gland. Because intake of radioactive iodine is relatively constant with age, radiation doses are usually greater in smaller children. Both factors are considered in determining the multipliers given in the age panel of Table 2. A proportion of exposure came from sources other than milk (often as much as one-third of the total) so a different multiplier should be assigned to the non-milk drinker. Persons who consumed more milk than the average are allocated a multiplier of __. Finally, a gender-specific factor is included because females are more sensitive to cancer induction by a factor of about __.

Step 4. Determine individual relative risk.

Procedure: Multiply the 4 factors obtained from Steps 2 and 3 to determine your relative risk value.

Explanation: The product of the 4 factors is a number that indicates the risk of developing thyroid abnormalities (cancer in particular) as a result of exposure to I-131 from the nuclear weapons testing. This value is risk relative to the group of males aged 5-14 and above at the time of the testing. Men in that age group are the "reference category" because, of all the age and gender combinations, they represent the group with the lowest risk of exposure-induced disease (older men were of minimal, if not zero, risk). The risk of those in the reference category is about ___ on the scale developed here. The risk anyone else might calculate could be as high as ____. Example calculations are provided to assist the reader.

Here are some examples of the way the calculations would work.

•   Male, born in 1941, living in county "a" (medium-deposition county) who drank commercial milk at twice or more the average rate in the 1950s:

From Table 1A, column 2	CO_M
From Table 2, age section	A_{10-19}
From Table 2, milk consumption section	CR_{2x}
From Table 2, gender section	G_M
Risk (relative to reference group)	$= CO_M \times A_{10-19} \times CR_{2x} \times G_M$

•   Female, born in 1951, living in county "b" (high-deposition county) who drank milk from a backyard goat at the normal rate:

From Table 1B, column 3	GM_H
From Table 2, age section	A_{0-9}
From Table 2, milk consumption section	CR_{avg}
From Table 2, gender section	GF_M
Risk (relative to reference group)	$= GM_h \times A_{0-9} \times CR_{avg} \times GF_M$

Table 3 provides information for the person making the assessment for understanding the calculated risk value.

TABLE 3 Interpreting the Results

SCORE[a]	WHAT IT MEANS	WHAT YOU SHOULD DO
Less than ———	For all practical purposes, you have no risk of radiation related thyroid cancer. Your lifetime risk of developing thyroid cancer is the same as for people not exposed to radiation, about 1 in 400 for males, and 1 in 150 for females.[b]	Nothing.
_____ to _____	You have been exposed to a risk of very small magnitude; when compared with many other risks, it is of little consequence. Your lifetime risk of developing radiation related thyroid cancer is equal to an increase of a few percent over the background rate of thyroid cancer. This would mean your lifetime risk would be about 1 in _____ for males, and about 1 in _____ for females.	Nothing.
_____ to _____	You have been exposed to a moderately increased risk of thyroid cancer. Your lifetime risk of radiation related thyroid cancer is about twice that of people not exposed to radiation, or about 1 in _____ for males, and 1 in _____ for females.	If you are concerned, read the enclosed section of this brochure about thyroid cancer.[c]
_____ to _____	Your risk of thyroid cancer could be significantly greater than normal. Your lifetime risk of radiation related thyroid cancer could be within the following range: on the low side it would be 1 in _____ for males, and 1 in _____ for females; on the high side it could be as much as 1 in _____ for males or 1 in _____ for females.	If you are concerned, read the enclosed section of this brochure about thyroid cancer.[c]

[a]The committee has discussed the issues of public numeracy and suggests that a useful scale would be a range from less than 1.0 to 100.

[b]Lifetime risk for unexposed persons is discussed in Chapter 3 of this report.

[c]This committee recommends that DHHS develop suitable written materials as part of this model Web page or brochure that would provide information to the person about the nature and course of thyroid cancer and the possible risks and benefits of testing for thyroid disease. The written material might use some of the information provided in this report in the background paper (Appendix F) and from the section of Chapter 4 that addresses the benefits and harms of screening. Information concerning thyroid cancer is available on numerous Web sites including those of the NCI, the American Cancer Society, the Hanford Health Education Network, and the American Thyroid Association.

6

Research Needs

Thyroid cancer is one of the better studied malignancies associated with exposure to ionizing radiation, but there are still many gaps and shortcomings in the data currently available. These deficiencies impinge on our ability to estimate risk appropriately, to delineate those factors that modify risk (such as age at exposure), to understand the biologic processes that underlie the occurrence of and prognosis for radiation-related thyroid cancer, to distinguish between those thyroid nodules that ultimately will be harmful to the patient from those that are not, and to communicate to members of the public and health care professionals the risk that results from exposure to ionizing radiation. The Committee on Exposure of the American People to I-131 from Nevada Atomic-Bomb Tests has examined the more salient of these needs and sets out below a series of recommendations for research. These recommendations are by no means exhaustive, but they attempt to address the more pressing issues the U.S. Department of Health and Human Services (DHHS) must confront. For convenience, these recommendations and their rationales are set out under four rubrics: epidemiology, the biology of radiation-related thyroid cancer, clinical practice, and risk communication.

Though this committee does not believe that further research into understanding all of the impacts of global fallout on the continental United States will make an important contribution to current public health, there are those that might desire a more complete understanding of this matter. Presently, sufficient information does not exist to easily evaluate the possible increase in disease rates in the United States due to exposure of the public to fallout of other radionuclides from U.S. tests or global fallout from tests conducted in other countries. For such an evaluation, further reconstruction of deposition records, transfer of radionuclides in the environment, and calculations to support various exposure scenarios would be needed. Such an effort would easily be on the scale or greater of that which NCI accomplished in support of its 1997 report. This ought to only be considered as a research recommendation if the government or public believes attaining a fuller understanding of nuclear testing to be an issue that supercedes other more basic health issues.

EPIDEMIOLOGY

The report of the National Cancer Institute (NCI 1997a), together with the presentation material provided by Charles Land (Appendix B, this report), gives a theoretical basis for positing that nuclear-weapons testing at the Nevada Test Site led to exposures to the thyroid that resulted in significant excess risk of thyroid disease for some members of the population of the United States, depending on age, geography, and individual consumption and source of milk. As indicated in Land's analyses, and as further discussed in Chapters 2 and 3 of this report, there are still important uncertainties in the assessment of the size of the exposures and their influence on thyroid cancer risk. No epidemiologic evidence was presented either in the NCI (1997a) report or in concomitant releases of information that thyroid cancer risk differs significantly by birth cohort, geography, or behavior as a consequence of the iodine-131 fallout from weapons testing. It would be premature in this committee's view to initiate changes in medical practice based solely on the theoretical calculations unless there is also empirical evidence of an increased risk. Accordingly, a comprehensive study of past incidence of thyroid disease—using the full range of resources from existing tumor registries, including but not restricted to those associated with the Surveillance, Epidemiology, and End Results (SEER) program—is clearly important and essential to more realistic risk estimates. Such studies have the statistical power to detect increases in risk of the size predicted by Land, by birth cohort, and possibly by geography.

Studies of current incidence (case-control studies) are feasible and would be useful in evaluating whether behavioral factors such as milk consumption in the 1950s are indeed related to individual risk. For example, a study of 500-1,000 cases matched to an equal number of controls would have good statistical power to detect increases due to behavioral factors (milk consumption), of 50 percent or more. Such a difference in risk would correspond to a difference of 0.08 Gy in dose using the Ron and others (1995) risk estimates for those exposed as very young children. The most heavily exposed birth cohort (persons exposed between the ages of 0 and 4 in 1952) consists of approximately 18 million people. At an average current rate of thyroid cancer of 11 per 100,000 person-years (assuming a 40 percent increased risk due to exposure), this entire cohort would produce about 2,000 cases of thyroid cancer in a year. Case-control studies, however, are likely to be subject to recall biases—especially given the degree of publicity the release of the NCI report received. Nevertheless, case-control studies would appear to be the only reasonable way of empirically relating behavior to risk.

Therefore, this committee recommends that NCI consider the feasibility of such studies carefully, with attention given in study design and analysis to minimizing or controlling recall biases to the extent possible.

The need for more epidemiologic data on the effects of exposure of young people to I-131 is acute. Weapons testing took place in several other parts of the world, in some cases with even greater fission yields than in Nevada. The NCI's dose reconstruction for Nevada is by far the most comprehensive for such situations to date, and it reveals the great range at which doses to the thyroid gland of public health significance can apply to specific groups in the population. Although the likelihood of a resumption of atmospheric testing is remote, accidents to nuclear facilities also can release substantial quantities of I-131, as demonstrated in 1986 by the Chernobyl reactor accident. The dramatic and early increase in cancer of the thyroid among individuals exposed in childhood after this accident was clearly visible because of the very low spontaneous incidence of the disease. As the exposed population ages and as spontaneous incidence rates increase, the task of identifying the proportion of thyroid disease attributable to I-131 exposure will become increasingly difficult. Nevertheless, the Chernobyl experience provides a unique opportunity to learn for the future and gain further information relevant to the public-health consequences of the Nevada testing.

A recent editorial (Baverstock 1998) drew attention to the need for a coherent international response to ensure that the opportunity to learn from this experience is not lost. And in this regard, this committee urges continued support of the collaborative study involving NCI and the Ukrainian Research Center for Radiation Medicine of thyroid cancer among Ukrainian children exposed to the Chernobyl accident.

BIOLOGY OF RADIATION-INDUCED THYROID CANCER

Several questions of significance remain with respect to the biology of radiation-induced thyroid cancer. The first and most important is that of the relative biological effectiveness (RBE) of I-131 compared with external exposure. There is a discrepancy between previously held thoughts on the RBE of I-131 and the data from Chernobyl, which suggest an RBE closer to 1. Experimental studies have demonstrated no difference in effectiveness between x-rays and I-131 (Lee and others 1982). It is appropriate to consider performing further experimental studies to address the question of x-ray versus I-131 RBE more compellingly.

A second question is that of whether there is any difference between the biologic behavior of radiation-induced versus spontaneous thyroid cancer. Some reports suggest that radiation-induced lesions are more aggressive and malignant, but it is not clear how well-founded these reports are. Further studies of the relative malignancy of radiation-related and spontaneous thyroid neoplasms will be important in understanding the true risk for exposed populations.

Recent evidence provided by Jhiang and others (1996) shows that *ret*/PTC1 (papillary thyroid cancer 1) is a genetic event leading by itself to the development of thyroid carcinoma in transgenic mice, an observation that has been confirmed (Santoro and others 1996). Those observations provide strong evidence for the role of *ret* rearrangements in the development of papillary thyroid cancer, which should be considered in recommendations for future research in patients exposed to radiation from atomic weapons fallout. Moreover, such research could have an important bearing on the nature of the dose-response relationship for thyroid cancer. If, for example, the *ret* oncogene is commonly activated by a paracentric inversion, a two-hit chromosomal phenomenon, this would suggest a probable linear-quadratic dose-response relationship because paracentric inversions become more likely at higher doses where two hits increase in frequency.

CLINICAL PRACTICE

As set forth in Chapter 3, the introduction of sophisticated diagnostic tests, such as ultrasonography, has resulted in the discovery of many microcancers that are unlikely to harm the patient. A major challenge for medical research is to differentiate those microcancers that are clinically significant from those that will never harm the patient. Failure to make this differentiation will result in some patients undergoing treatment for harmless diseases and others having their diseases ignored imprudently. This problem of microcancers is not unique to the thyroid gland; they are found even more commonly in the breast and prostate. The committee therefore recommends support of clinical studies designed to achieve a better means of differentiating between potentially harmful and harmless microcancers.

RISK COMMUNICATION

Studies of how to communicate risk appropriately are far more meager than are studies of the risks themselves. Agencies and policy makers historically have been more willing to acknowledge the role of expertise in questions about scientific process than in questions about communication. Thus, the communication advice offered in this document is grounded in fewer studies than are the deliberations about the risk of thyroid cancer or the process and prognosis of that disease.

DHHS would be well served by efforts to understand better how individuals make judgments about risk situations and, more specifically, how they use information to do so. Particularly relevant to the thyroid cancer situation are the following questions:

• What analytical skills do individuals bring to information about frequency and magnitude of risk?

- How can communicators offer explanations of risk concepts and processes that convey the intended meanings?
- How do individuals make decisions about whom they trust when seeking information about risks?
- How should risks be communicated so that affected audiences are likely to interpret the information offered as being personally relevant?
- When an information source has lost its credibility, how does that source legitimately regain trust?

Given the significant uncertainties surrounding I-131 exposure and cancer risk, the committee suggests that DHHS consider research to develop a better understanding of how people perceive the benefits and harms of cancer screening and how those perceptions are affected by different ways of presenting quantitative information and different ways of structuring clinician-patient communication. In addition to studies not specific to thyroid cancer or radiation-exposed populations, the evaluation component of the Agency for Toxic Substances and Disease Registry (ATSDR) medical-monitoring program might consider the feasibility of a controlled study to compare different information formats or different counseling strategies for patients who come in for testing.

As envisaged, the ATSDR medical-monitoring program is not likely to produce useful information about mortality effects of screening for thyroid cancer because people who are screened would be self-selected, and high long-term survival rates can be expected without screening. The program's reports on positive and negative screening results and subsequent follow-up tests, procedures, and findings should still be tracked. In considering its research priorities, DHHS might also consider an outside evaluation of patient experience and perceptions of harms and other benefits of the ATSDR program, although it is not clear that the results could be generalized.

References

AACE (American Association of Clinical Endocrinologists). 1997a. AACE clinical practice guidelines for the diagnosis and management of thyroid nodules. *Web Site* Thyroid Nodule Guidelines. Available at: http://www2.nsysu.edu.tw/hclam/nodule.html.

AACE (American Association of Clinical Endocrinologists). 1997b. Americans urged to stick their necks out to detect common cancer that is often overlooked: New self exam and clinical guidelines issues to improve detection and treatment of thyroid cancer. Available at: http://www.aace.com/guidelines/taprelease.html.

ACS (American Cancer Society). 1998a. Cancer facts and figures 1998: graphical data. Available at: http://www.cancer.org/cidSpecificCancers/thyroid/early.html.

ACS (American Cancer Society). 1998b. Cancer facts and figures 1998: graphical data. Available at: http://www.cancer.org/cidSpecificCancers/thyroid/early.html.

Ashcraft MW, Van Herle AJ. 1981. Management of thyroid nodules. II: Screening techniques, thyroid suppressive therapy, and fine needle aspiration. Head and Neck Surgery 3(4):197-322.

ATA (American Thyroid Association). 1997. Guidelines for physicians. Available at: http://www.thyroid.org/members/guideline.html.

ATSDR. (Agency for Toxic Substances and Disease Registry). 1997. Hanford Medical Monitoring Program: Background Consideration Document and ATSDR Decision. Pub. No. PB97-193072. Atlanta, GA: U.S. Department of Health and Human Services.

Baes CF, Orton TH. 1979. Productivity of argricultural crops and forage, Y_v. In: Hoffman FO, Baes CF (eds.), A Statistical Analysis of Selected Parameters for Predicting Food Chain Transport and Internal Doses of Radionuclides. Washington, DC: U.S. Nuclear Regulatory Commission. pp. 15-29.

Banta HD, Engel GL, Schersten T. 1992. Volume and outcome of organ transplantation. International Journal of the Technological Assessment of Health Care 8(3):490-505.

Barry MJ, Fowler Jr FJ, Mulley Jr AG, Henderson Jr JV, Wennberg JE. 1995. Patient reactions to a program designed to facilitate patient participation in treatment decisions for benign prostatic hyperplasia. Medical Care 33:771-782.

Baverstock KF. 1998. Chernobyl and public health. British Medical Journal 316:952-953.

Beck HA, Helfer IK, Bouville A, Dreicer M. 1990. Estimates of fallout in the continental U.S. from Nevada weapons testing based on gummed-film monitoring data. Health Physics 59(5):565-576.

Beck JR, Kattan MW, Miles BJ. 1994. A critique of the decision analysis for clinically localized prostate cancer. Journal of Urology 152(5, Pt 2):1894-1899.

Becker DV, Robbins J, Beebe GW, Bouville AC, Wachholz BW. 1996. Childhood thyroid cancer following the Chernobyl accident—A status report. Endocrinology and Metabolism Clinics of North America 25:197-211.

Becker MH, Joseph JG. 1988. AIDS and behavioral change to reduce risk: A review. American Journal of Public Health 78:394-410.

Begley S. 1997. The mammogram war. Newsweek, February 24, 1997. pp. 54-58.

Belfiore A, Rosa GL, Giuffrida D, Regalbuto C, Lupo L, Fiumara A, Ippolito O. 1995. The management of thyroid nodules. Journal of Endocrinological Investigation 18(2):155-158.

BIOMOVS. 1991. Multiple model testing using Chernobyl fallout data of I-131 in forage and milk and Cs-137 in forage, milk, beef, and grain. Technical Report 13, Part 1.

Boey J, Hsu C, Collins RJ. 1986. False negative errors in fine-needle aspiration biopsy of dominant thyroid nodules: A prospective follow-up study. World Journal of Surgery 10:623.

Boiko PE, Morrill RL, Flynn J, Faustman EM, Van Belle G, Omenn GS. 1996. Who holds the stakes? A case study of stakeholder identification at two nuclear weapons production sites. Risk Analysis 16:237-249.

Bongarzone I, Fugazzola L, Vigneri P, Mariani L, Mondellini P, Pacini F, Basolo F, Pinchera A, Pilotti S, Pierotti MA. 1996. Age-related activation of the tyrosine kinase receptor proto-oncogene RET and NTRK1 in papillary thyroid carcinoma. Journal of Clinical Endocrinology and Metabolism 81:2006-2009.

Brander A, Viikinkiski P, Tuuhea P, Voutilainen L, Kivisaari L. 1992. Clinical versus ultrasound examination of the thyroid gland in common clinical practice. Journal of Clinical Ultrasound 20(1):37-42.

Brown J, Goossens LHJ, Draan BCP, et al. 1997. Probabilistic Accident Consequence Uncertainty Analysis: Food Chain Uncertainty Assessment. Volumes 1 and 2, NUREG/CR-6523. Washington, DC: Nuclear Regulatory Commission/Commission of European Communities.

Cady B, Rossi R. 1988. An expanded view of risk-group definition in differentiated thyroid carcinoma. Surgery 104:947-953.

Campbell JE, Murphy GK, Goldin AS. 1959. The occurrence of strontium-90, iodine-131, and other radionuclides in milk. American Journal of Public Health 49:186-196.

Carroll BA. 1982. Asymptomatic thyroid nodules: Incidental sonographic detection. American Journal of Roentgenology 138(3):499-501.

CDC (Centers for Disease Control and Prevention). 1996. CDC Guidelines: Improving the Quality. Atlanta, GA: Centers for Disease Control and Prevention.

Chamberlain AC. 1970. Interception and retention of radioactive aerosols by vegetation. Atmospheric Environment 4:57-78.

Charboneau JW, Hay ID, Mazzaferri EL. 1998. Perspective on Benign and Malignant Thyroid Nodules. In: Committee on Guidelines for Thyroid Cancer Screening. Washington, DC: Institute of Medicine.

Chen H, Zeiger MA, Gordon TA, Udelsman R. 1996. Parathyroidectomy in Maryland: Effects of an endocrine center. Surgery 120(6):948-952.

Church BW, Wheeler DL, Campbell CM, Nutley RV, Anspaugh LR. 1990. Overview of the Department of Energy's off-site radiation exposure review project (ORERP). Health Physics 59:503-510.

CommerceNet. 1997. Startling increase in Internet shopping reported in CommerceNet/Nielsen Media Research Study. Available at: http://www.commerce.net/work/pilot/Nielsen_96/press_97/html.

Cooper DS. 1991. Treatment of thyrotoxicosis. In: Braverman LE, Utiger RD (eds.), Werner and Ingbar's, The Thyroid. Philadelphia: J.B. Lippincott Co. pp. 887-916.

Cronkite EP, Conard RA, Bond VP. 1997. Historical events associated with fallout from Bravo shot—Operation Castle and 25 y of medical findings. Health Physics 73(1):176-186.

CTFPHE (Canadian Task Force on the Periodic Health Examination). 1994. Canadian Guide to Clinical Preventive Health Care. Canada Communications Group.

Davison WP. 1983. The third-person effect in communication. Public Opinion Quarterly 47:1-15.

De Keyser LFM, Van Herle AJ. 1985. Differentiated thyroid cancer in children. Head Neck Cancer 8:100-114.

De Vathaire F, Fragu P, Francois P, Benhamou S, Ward P, Benhamou E, Avril MF, Grimaud E, Sancho-Garnier H, Parmentier C. 1993. Long-term effects on the thyroid of irradiation for skin angiomas in childhood. Radiation Research 133:381-386.

DeGroot LJ. 1988. Radiation and thyroid disease. Baillieres Clinical Endocrinology and Metabolism 2:777-791.

Demeure MJ, Clark OH. 1990. Surgery in the treatment of thyroid cancer. Endocrinology and Metabolism Clinics of North America 19(3):663-683.

Derwin B. 1981. Audience as listener and learner, teacher and confidante: The sense-making approach. In: Rice RE, Atkins CK (eds.), Public Communication Campaigns. 2nd ed. Newbury Park: Sage. pp. 76-86.

Dinneen SF, Valimaki MJ, Bergstrahl EJ, Goellner JR, Gorman CA, Hay ID. 1995. Distant metastases in papillary thyroid carcinoma: 100 cases observed at one institution during 5 decades. Journal of Clinical Endocrinology and Metabolism 80:2041-2045.

DOE (Department of Energy). 1994. United States Nuclear Tests, July 1945 through September 1992. DOE/NV-209 (Rev. 14). Las Vegas, NV: U.S. Department of Energy.

Douglas M, Wildavsky A. 1982. Risk and Culture. Berkeley, CA: University of California Press.

Dunwoody S, Neuwirth K. 1991. Coming to terms with the impact of communication on scientific and technological risk judgments. In: Wilkins L, Patterson P (eds.), Risky Business. New York: Greenwood Press. pp. 11-30.

Dunwoody S, Peters HP. 1992. Mass media coverage of technological and environmental risks: A survey of research in the United States and Germany. Public Understanding of Science 1:199-230.

Durbin PW, Lynch J, Murray S. 1970. Average milk and mineral intakes (calcium, phosphorus, sodium and potassium) of infants in the U.S. from 1954 to 1968. Implications for estimating annual intakes of radionuclides. Health Physics 19:187-222.

Eddy DM. 1991. In: Common Screening Tests. Philadelphia, PA: American College of Physicians.

Eddy DM. 1997. Breast cancer screening in women younger than 50 years of age: what's next? Annals of Internal Medicine 127(11):1035-1036.

Emanuel EJ, Emanuel LL. 1992. Four models of the physician-patient relationship. Journal of the American Medical Association 267:2221-2226.

Emerick GT, Duh QY, Siperstein AE, Burrow GN, Clark OH. 1993. Diagnosis, treatment, and outcome of follicular thyroid carcinoma. Cancer 72:3287-3295.

Ende J, Kazis L, Ash A, Moskowitz M. 1989. Measuring patients' desire for autonomy: Decision making and information-seeking preferences among medical patients. Journal of General Internal Medicine 4:23-30.

England TR, Rider BF. 1994. Evaluation and Compilation of Fission Product Yields-1993. LA-UR-94-3106 Los Alamos, NM: Los Alamos National Laboratory (also available as ENDF-349, National Nuclear Data Center, Upton, NY: Brookhaven National Laboratory).

EPA (Environmental Protection Agency). 1984. Offsite Environmental Monitoring Report. EPA-600/4-84-040. Las Vegas, NV: Environmental Protection Agency, Environmental Monitoring Systems Laboratory.

Ezzat S, Sarti DA, Cain DR, Braunstein GD. 1994. Thyroid incidentalomas: Prevalence by palpation and ultrasonography. Archives of Internal Medicine 154:1838-1840.

Fagin JA. 1994a. Molecular genetics of human thyroid neoplasms. Annual Review of Medicine 45:45-52.

Fagin JA. 1994b. Molecular pathogenesis of human thyroid neoplasms. Thyroid Today 17:1-7.

Falk SA. 1997. Complications of thyroid surgery: An overview. In: Falk SA (eds.), Thyroid Disease: Endocrinology, Surgery, Nuclear Medicine, and Radiotherapy. Second Edition. Philadelphia, PA: Lippincott-Raven. pp. 697-704.

Farid NR, Shi Y, Zou M. 1994. Molecular basis of thyroid cancer. Endocrinology Review 15:202-232.

Fisher JD, Fisher WA. 1992. Changing AIDS-risk behavior. Psychological Bulletin 111:455-474.

Fishkin JS. 1996. The televised deliberative poll: An experiment in democracy. Annals of the American Academy of Political and Social Science 546:132-140.

Fleming ID, Black TL, Thompson EI, Pratt C, Rao B, Hustu O. 1985. Thyroid dysfunction and neoplasia in children receiving neck irradiation for cancer. Cancer 55:1190-1194.

Flood AB, Wennberg JE, Nease Jr RF, Fowler Jr FJ, Ding J, Hynes LM. 1996. The importance of patient preference in the decision to screen for prostate cancer. Journal of General Internal Medicine 11(6):377-378.

Freimuth V. 1998. Science on demand before the Web: The cancer hotline. In: Paper presented at annual meeting of the American Association for the Advancement of Science.

Freudenberg WR. 1988. Perceived risk, real risk: Social science and the art of probabilistic risk assessment. Science 242:44-49.

Freudenberg WR. 1992. Heuristics, biases, and not-so-general publics: expertise and error in the assessment of risks. In: Krimsky S, Golding D (eds.), Social Theories of Risk. Westport, CT: Prager. pp. 229-249.

Freudenberg WR, Rursch RA. 1994. The risks of "putting the numbers in context," a cautionary tale. Risk Analysis 14:949-958.

Fugazzola L, Pierotti MA, Vigano E, Pacini F, Vorontsova TV, Bongarzone I. 1996. Molecular and biochemical analysis of RET/PTC4 a novel oncogenic rearrangement between RET and ELE1 genes, in a post-Chernobyl papillary thyroid cancer. Oncogene 13:1093-1097.

Fugazzola L, Pilotti S, Pinchera A, Vorontsova TV, Mondellini P, Bongarzone I, Greco A, Astakhova L, Butti MG, Demidchik EP, Pacini F, Pierotti MA. 1995. Oncogenic rearrangements of the RET proto-oncogene in papillary thyroid carcinomas from children exposed to the Chernobyl nuclear accident. Cancer Research 55:5617-5620.

Fujiwara S, Carter RL, Akiyama M, Akahoshi M, Kodama K, Shimaoka K, Yamakido M. 1994. Autoantibodies and immunoglobulins among atomic bomb survivors. Radiation Research 137:89-95.

Fuscoe JC, Zimmerman LJ, Fekete A, Setzer RW, Rossiter BJ. 1992. Analysis of X-ray-induced HPRT mutations in CHO cells: insertion and deletions. Mutation Research 269:171-183.

Garcia-Mayor RV, Perez Mendez LF, Paramo C, Luna Cano R, Rego Iracta A, Regal M. 1997. Fine-needle aspiration biopsy of thyroid nodules: impact on clinical practice. Journal of Endocrinological Investigation 20(8):482-487.

Gembicki M, Stozharov AN, et al. 1997. Iodine deficiency in Belarussian children as a possible factor stimulating the irradiation of the thyroid gland during the Chernobyl catastrophe. Environmental Health Perspectives 105(Supplement 6):1487-1491.

Gharib H. 1997. Changing concepts in the diagnosis and management of thyroid nodules. Endocrinology and Metabolism Clinics of North America 26(4):777-800.

Gharib H, Goellner JR. 1993. Fine-needle aspiration biopsy of the thyroid: an appraisal. Annals of Internal Medicine 118(4):282-289.

Gharib H, Mazzaferri EL. 1998. Thyroxine supressive therapy in patients with nodular thyroid disease. Annals of Internal Medicine 128(5):386-394.

Gigerenzer G. 1996. The psychology of good judgment: Frequency formats and simple algorithms. Medical Decision Making 16:270-280.

Gilovich TD. 1991. How We Know What Isn't So: The Fallibility of Human Reason in Everyday Life. New York, NY: The Free Press.

Goepfert H, Dichtel WJ, Samaan NA. 1984. Thyroid cancer in children and teenagers. Archives of Otolaryngology 110:72-75.

Goulko GM, Chepurny NI, et al. 1998. Thyroid dose and thyroid cancer incidence after the Chernobyl accident assessments from the Zhytomir region (Ukraine). Radiation and Environmental Biophysics 36:261-273.

Grieco M, Santoro M, Berlingieri MT, Melillo RM, Donghi R, Bongarzone I, Pierotti MA, Della Porta G, Fusco A, Vecchio G. 1990. PTC is a novel rearranged form of the ret proto-oncogene and is frequently detected in vivo in human thyroid papillary carcinomas. Cell 60:557-563.

Grumbach K, Anderson GM, Luft HS, Roos LL, Brook R. 1995. Regionalization of cardiac surgery in the United States and Canada. Geographic access, choice, and outcomes. Journal of the American Medical Association 275(10):758-759.

Haas S, Trujilla A. 1993. Fine needle aspiration of thyroid nodules in a rural setting. American Journal of Medicine 94(4):357-361.

Hall P, Mattsson A, Boice J. 1995. Thyroid cancer after diagnostic administration of iodine-131. Radiation Research 145:86-92.

Hall TL, Layfield LJ, Philippe A, Hosenthal DL. 1989. Sources of diagnostic error in fine needle aspiration of the thyroid. Cancer 63:718.

Hallquist A, Hardell L, Degerman A, Wingren G, Boquist L. 1994. Medical diagnostic and therapeutic ionizing radiation and the risk for thyroid cancer: A case-contol study. European Journal of Cancer Prevention 3:259-267.

Hamilton P, Chiacchierini R, Kaczmarek R. 1989. A follow-up study of persons who had iodine-131 and other diagnostic procedures during childhood and adolescence. Rockville, MD: CDRH-Food & Drug Administration.

Hamilton TE, Van Belle G, LoGerfo JP. 1987. Thyroid neoplasia in Marshall Islanders exposed to nuclear fallout. Journal of the American Medical Association 258:629-635.

Hancock SL, Cox RS, McDougall IR. 1991. Thyroid diseases after treatment of Hodgkin's disease. New England Journal of Medicine 325:599-605.

Harach HR, Franssila KO, Wasenius VM. 1985. Occult papillary carcinoma of the thyroid: A "normal" finding in Finland. A systematic autopsy study. Cancer 56(3):531-538.

Harley NH, Albert R, Shore R, Pasternack B. 1976. Follow-up study of patients treated by X-ray epilation for tinea capitis. Estimation of the dose to the thyroid and pituitary glands and other structures of the head and neck. Physics of Medicine and Biology 21:631-642.

Harness JK, Thompson NW, McLeod MK, et al. 1992. Differentiated thyroid carcinoma in children and adolescents. World Journal of Surgery 16:547-554.

Hay ID. 1990. Papillary thyroid carcinoma. Endocrinology and Metabolism Clinics of North America 19:545-576.

Helfand M, Redfern CC. 1998. Screening for thyroid disease: An update. Annals of Internal Medicine 129:144-158.

Helfand M, Redfern CC, Sox HC. 1998. Screening for thyroid disease. Annals of Internal Medicine 129:141-143.

Hicks HG. 1981. Results of Calculations of External Gamma Radiation Exposure Rates from Fallout and the Related Radionuclide Compositions. UCRL-53152, Parts 1-8. Livermore, CA: Lawrence Livermore National Laboratory.

Hicks HG. 1982. Calculation of the concentration of any radionuclide deposited on the ground by offsite fallout from a nuclear detonation. Health Physics 42:585-600.

Hildreth NG, Schneider AB, Cave Jr WT. 1987. A comparative study between individuals receiving thymic irradiation in infancy and their nontreated siblings: Clinical and laboratory thyroid abnormalities. Radiation Research 110:458-467.

Hoffman DL, Novak TP. 1998. Bridging the racial divide on the Internet. Science 280:390-391.

Hoffman FO, Frank ML, Blaylock BG, Von Bernuth RD, Deming EJ, Graham RV, Mosbacher DA, Waters AE. 1989. Pasture grass interception and retention of ^{131}I, ^{7}Be and insoluble microspheres deposited in rain. ORNL-6542//Env. Sci. Div. Publ. No. 3247. Oak Ridge National Laboratory/Environmental Sciences Division.

Hoffman FO, Thiessen KM, Frank ML, Blaylock GB. 1992. Determining the collection efficiency of gummed paper for the deposition of radioactive contaminants in simulated rain. Health Physics 62(5):439-442.

Horlocker TT, Jay JE. 1985. Prevalence of incidental nodular thyroid disease detected during high-resolution parathyroid ultrasonography. In: Medeiros-Neto G, Gaitan E (eds.), Frontiers in Thyroidology. Volume 2. New York: Plenum Medical. pp. 1309-1312.

Horton RE. 1919. Rainfall interception. Monthly Weather Review 49:603.

Howard JE, Vaswanik A, Heotis P. 1997. Thyroid disease among the Rongelap and Utirik population-an update. Health Physics 73(1):190-198.

Hsiao YL, Chang TC. 1994. Ultrasound evaluation of thyroid abnormalities and volume in Chinese adults without palpable thyroid glands. Journal of the Formosan Medical Association 93(2):140-144.

Huff D. 1954. How to Lie with Statistics. New York: Norton.

Hughes RG, Hunt SS, Luft HS. 1987. Effects of surgeon volume and hospital volume on quality of care in hospitals. Medical Care 25(6):489-503.

Hung W. 1994. Well-differentiated thyroid carcinomas in children and adolescents. The Endocrinologist 4:117-126.

Hux JE, Naylor CD. 1995. Communicating the benefits of chronic preventive therapy: Does the format of efficacy data determine patients' preferences for therapeutic outcomes. Medical Decision Making 15:152-157.

Huysmans DAKC, Hermus ARMM, Edelbroek MAL, Tjabbes T, Oostdijk A, Ross HA, Corstens FHM, Kloppenborg PWC. 1997. Autoimmune hyperthyroidism occurring late after radioiodine treatment for volume reduction of large multinodular goiters. Thyroid 7:535-539.

Hyman HH, Sheatsley PB. 1974. Some reasons why information campaigns fail. Public Opinion Quarterly 11:412-423.

IAEA (International Atomic Energy Agency). 1996. Modelling of Radionuclide Interception and Loss Processess in Vegetation and of Transfer in Semi-Natural Ecosystems, Second Report of VAMP Terrestrial Working Group. IAEA-TECDOC-857. Vienna, Austria: International Atomic Energy Agency.

Imperato PJ, Nenner RP, Starr HA, Will TO, Rosenberg CR, Dearie MB. 1996. The effects of regionalization on clinical outcomes for a high risk surgical procedure: A study of the Whipple procedure in New York state. American Journal of Medical Quality 11(40):193-197.

Inskip P, Ekbom A, Galanti M, Grimelius L, Boice J. 1995. Medical diagnostic X-rays and thyroid cancer. Journal of the National Cancer Institute 87:1613-1621.

IOM (Institute of Medicine). Field MJ, Lohr KN. 1992. Guidelines for Clinical Practice. Washington, DC: National Academy Press.

Isaaks EH, Srivastava RM. An Introduction to Applied Geostatistics. 1989. New York, Oxford: Oxford University Press.

Ishida T, Izuo M, Ogawa T, Kurebayashi J, Satoh K. 1988. Evaluation of mass screening for thyroid cancer. Japanese Journal of Clinical Oncology 18:289-295.

Ito C, Kato M, Mito K, Ishibashi S, Matsumoto Y. 1987. Study of the effect of atomic bomb radiation on thyroid function. Hiroshima Journal of Medical Sciences 36:13-24.

Ito M, Yamashita S, Ashizawa K, Namba H, Hoshi M, Shibata Y, Sekine I, Nagataki S, Shigematsu I. 1995. Childhood thyroid diseases around Chernobyl evaluated by ultrasound examination and fine needle aspiration cytology. Thyroid 5:365-368.

Ito T, Seyama T, Iwamoto KS, Mizuno T, Tronko ND, Komissarenko IV, Cherstovoy ED, Satow Y, Takeichi N, Dohi K, Akiyama M. 1994. Activated RET oncogene in thyroid cancers of children from areas contaminated by Chernobyl accident. Lancet 344:259.

Jacob P, Goulko G, Heidenreich WF, Likhtarev I, Kairo I, Tronko ND, Bogdanova TI, Kenigsberg J, Buglova E, Drozdovitch V, Golovneva A, Demidchik EP, Balanov M, Zvonova I, Beral V. 1998. Thyroid cancer risk to children calculated. Nature 392:31-32.

Jarlov AE, Hegedus L, Gjorup T, Hansen JM. 1991. Observer variation in ultrasound assessment of the thyroid gland. Journal of Internal Medicine 229(2):159-161.

Jarlov AE, Nygard B, Hegedus L, Karstrup S, Hansen JM. 1993. Observer variation in ultrasound assessment of the thyroid gland. British Journal of Radiology 66(787):625-627.

Jhiang SM, Mazzaferri EL. 1994. The RET/PTC oncogene in papillary thyroid carcinoma. Journal of Laboratory and Clinical Medicine 123:331-337.

Jhiang SM, Sagartz JE, Tong Q, Parker-Thornburg J, Capen CC, Cho JY, Xing SH, Ledent C. 1996. Targeted expression of the RET/PTC1 oncogene induces papillary thyroid carcinomas. Endocrinology 137:375-378.

Jhiang SM, Smanik PA, Mazzaferri EL. 1994. Development of a single-step duplex RT-PCR detecting different forms of ret activation, and identification of the third form of in vivo ret activation in human papillary thyroid carcinoma. Cancer Letter 78:69-76.

Johnson RJ, Tversky A. 1984. Representations of perceptions of risk. Journal of Experimental Psychology: General 113:55-70.

Jungermann H, Pfister HR, Fischer K. 1996. Credibility, information preferences, and information interests. Risk Analysis 16:251-261.

Kahneman D, Tversky A. 1979. Prospect theory: An analysis of decision under risk. Econometrica 47:263-292.

Kasatkina EP, Shilin DE, Rosenbloom AL, Pykov MI, Ibragimova GV, Sokolovskaya VN, Matkovskaya AN, Volkova TN, Odoud EA, Bronshtein MI, Poverenny AM, Mursankoba NM. 1997. Effects of low level radiation from the Chernobyl accident in a population with iodine deficiency. European Journal of Pediatrics 156:916-920.

Kaspar JF, Mulley AG, Wennberg JE. 1992. Developing shared decision-making programs to improve the quality of health care. Quality Review Bulletin 18:183-190.

Kasperson RE, Golding D, Tuler S. 1992. Social distrust as a factor in siting hazardous facilities and communicating risks. Journal of Social Issues 48:161-187.

Kassirer JP, Pauker SG. 1981. The toss-up. New England Journal of Medicine 305:1467-1469.

Katayama S, Shimaoka K, Piver MS, Osman G, Tsukada Y, Suh O. 1985. Radiation associated hyper-thyroidism in patients with gynecological malignacies. Journal of Medicine 16:587-596.

Kay TWH, Heyma P, Harrison LC, Martin FIR. 1987. Graves disease induced by radioactive iodine. Annals of Internal Medicine 107:857-858.

Kazakov VS, Demidchik EP, et al. 1992. Thyroid cancer after Chernobyl (letter). Nature 359(6390):21.

Kerber RA, Till JE, Simon SL, Lyon JL, Thomas DC, Preston-Martin S, Rallison ML, Lloyd RD, Stevens W. 1993. A cohort study of thyroid disease in relation to fallout from nuclear weapons testing. Journal of the American Medical Association 270:2076-2082.

Klugbauer S, Demidchik EP, Lengfelder E, Rabes HM. 1998. Detection of a novel type of RET rearrangement (PTC5) in thyroid carcinomas after Chernobyl and analysis of the involved RET-fused gene RFG5. Cancer Research 58:198-203.

Klugbauer S, Lengfelder E, Demidchik EP, Rabes HM. 1995. High prevalence of RET arrangement in thyroid tumors of children from Belarus after the Chernobyl reactor accident. Oncogene 11:2459-2467.

Klugbauer S, Lengfelder E, Demidchik EP, Rabes HM. 1996. A new form of RET rearrangement in thyroid carcinomas of children after the Chernobyl reactor accident. Oncogene 13:1099-1102.

Knapp HA. 1963. Iodine-131 in Fresh Milk and Human Thyroids Following a Single Deposition of Nuclear Test Fallout. TID-19266 (Health and Safety, TID-4500, 24th Ed.). Washington, DC: Atomic Energy Commission, Divison of Biology and Medicine, Fallout Studies Branch.

Larsen PR, Conard RA, Knudsen KD, Robbins J, Wolff J, Rall JE, Nicoloff JT, Dobyns BM. 1982. Thyroid hypofunction after exposure to fallout from a hydrogen bomb explosion. Journal of the American Medical Association 247:1571-1575.

Leard LE, Savides TJ, Ganiats TG. 1997. Patient preference for colorectal cancer screening. Journal of Family Practice 45(3):211-218.

Lee W, Chiacchierini RP, Schlein B, et al. 1982. Thyroid tumors following I-131 or localize X-irradiation to the thyroid and pituitary glands in rats. Radiation Research 92:307-319.

Lee W, Youmans H. 1970. Doses to the central nervous system of children resulting from X-ray therapy for tinea capitis. BRH/DBE 70-4. Rockville, MD: Bureau of Radiological Health, FDA.

Lessard E, Miltenberger R, Conard R, et al. 1985. Thyroid dose for people at Rongelap, Utirik and Sifo on March 1, 1954. BNL-51882. UC-48. Upton, NY: Brookhaven National Laboratory, Safety and Environmental Protection Division.

Likhtarev IA, Gouldon GM, et al. 1995. Evaluation of the I-131 thyroid-monitoring measurements performed in Ukraine during May and June of 1986. Health Physics 69:6-15.

Lin JD, Huang BY. 1997. Thyroid ultrasonography with fine-needle aspiration cytology for the diag-nosis of thyroid cancer. Journal of Clinical Ultrasound 25(3):111-118.

Lindberg S, Karlsson P, Arvidsson B, Holmberg E, Lundberg L, Wallgren A. 1995. Cancer incidence after radiotherapy for skin hemangioma during infancy. Acta Oncologica 34:735-740.

Lindsay S, Chaikoff IL. 1964. The effects of radiation on the thyroid gland with particular reference to the induction of thyroid neoplasms: A review. Cancer Research 24:1099-1107.

Litwin MS. 1994. Measuring health related quality of life in men with prostate cancer. Journal of Urology 152(5, Pt 2):1882-1887.

Liu JT, Smith VK. 1990. Risk communication and attitude change. Taiwan's national debate over nuclear power. Journal of Risk and Uncertainty 3:332-349.

LiVolsi VA. 1997. Pathology of thyroid disease. In: Falk SA (eds.), Thyroid Disease: Endocrinology, Surgery, Nuclear Medicine, and Radiotherapy. Second Edition. Philadelphia: Lippincott-Raven. pp. 65-104.

Loeffler JS, Tarbell NJ, Garber JR, Mauch P. 1988. The development of Grave's disease following radiation therapy in Hodgkin's disease. International Journal of Radiation Oncology, Biology, Physics 14:175-178.

Lopes LL. 1987. Between hope and fear: the psychology of risk. Experimental Social Psychology 20:255-295.

Lowery SA, DeFleur ML. 1995. Milestones in Mass Communication Research. 3rd. White Plains, NY: Longman.

Lundell G, Holm LE. 1980. Hypothyroidism following ^{131}I therapy for hyperthyroidism in relation to immunologic parameters. Acta Radiologica (Oncology) 19:449-454.

Lundell M, Hakulinen T, Lindell B, Holm LE. 1994. Thyroid cancer after radiotherapy for skin hemangioma in infancy. Radiation Research 140:334-339.

Maxon HR, Saenger EL. 1996. Biological effects of radioiodines on the human thyroid gland. In: Braverman LE, Utiger RD (eds.), Werner and Ingbar's, The Thyroid. Philadelphia, PA: Lippincott-Raven. pp. 342-351.

Mazur A. 1981. The Dynamics of Technical Controversy. Washington, DC: Communication Press.

Mazur DJ, Hickam DH. 1990a. Interpretation of graphic data by patients in a general medicine clinic. Journal of General Internal Medicine 5:402-405.

Mazur DJ, Hickam DH. 1990b. Treatment preferences of patients and physicians: Influences of summary data when framing effects are controlled. Medical Decision Making 10:2-5.

Mazzaferri EL. 1991. Carcinoma of follicular epithelium: Radioiodine and other treatments and outcomes. In: Braverman LE, Utiger RD (eds.), Werner and Ingbar's, The Thyroid. 6th ed. Philadelphia: J.B. Lippincott Co. pp. 1138-1165.

Mazzaferri EL. 1993a. Management of a solitary thyroid nodule. New England Journal of Medicine 328:553-559.

Mazzaferri EL. 1993b. Thyroid carcinoma: Papillary and follicular. In: Mazzaferri EL, Samaan N (eds.), Endocrine Tumors. Cambridge: Blackwell Scientific Publications Inc. pp. 278-333.

Mazzaferri EL. 1996. Radioactive and other treatments and outcomes. In: Braverman LE, Utiger RD (eds.), Werner and Ingbar's the Thyroid. 7th Edition. New York: Lippincott. pp. 922-945.

Mazzaferri EL, Jhiang SM. 1994. Long-term impact of initial surgical and medical therapy on papillary and follicular thyroid cancer. American Journal of Medicine 97:418-428.

Mazzaferri EL, Young RL, Oertel JE Kemmerer WT, Page CP. 1977. Papillary thyroid carcinoma: the impact of therapy in 576 patients. Medicine (Baltimore) 56(3):171-196.

McCormick KA, Moore SR, Siegel RA. 1994. Clinical Practice Guideline Development: Methodology Perspectives. AHCPR Pub. No. 95-0009. Washington, DC: U.S. Department of Health and Human Services.

McNeil BJ, Pauker SG, Sox HC, Tversky A. 1982. On the elicitation of preferences for alternative therapies. New England Journal of Medicine 306:1259-1262.

McTiernan A, Weiss N, Daling J. 1984. Incidence of thyroid cancer in women in relation to previous exposure to radiation therapy and history of thyroid disease. Journal of the National Cancer Institute 73:575-581.

Merchant WJ, Thomas SM. 1995. The role of thyroid fine needle aspiration (FNA) cytology in a district general hospital setting. Cytopathology 6(6):409-418.

Mettler Jr. FA, Williamson MR. 1992. Thyroid nodules in the population living around Chernobyl. Journal of the American Medical Association 268(5):616-619.

Moosa M, Mazzaferri EL. 1997. Occult thyroid carcinoma. Cancer Journal 10:180-188.

Morgan MW, Deber RB, Llewellyn-Thomas HA, Gladstone P, Cusimano RJ, O'Rourke K. 1997. A randomized trial of the ischemic heart disease shared decision making program: An evaluation of a decision aid [abstract]. Journal of General Internal Medicine 12:62.

Morimoto I, Yoshimoto Y, Sato K, Hamilton HB, Kawamoto S, Izumi M. 1987. Serum TSH, thyroglobulin, and thyroidal disorders in atomic bomb survivors exposed in youth: 30-year follow-up study. Journal of Nuclear Medicine 28:1115-1122.

Mortensen JD, Woolner LB, Bennett WA. 1955. Gross and microscopic findings in clinically normal thyroid glands. Journal of Clinical Endocrinology 15:1270.

Mutz DC. 1992. Impersonal influence: Effects of representations of public opinion on political attitudes. Political Behavior 14:89-122.

Nagataki S, Shibata Y, Inoue S, Yokoyama N, Izumi M, Shimaoka K. 1994. Thyroid diseases among atomic bomb survivors in Nagasaki. Journal of the American Medical Association 272:370.

National Research Council. 1989. Improving Risk Communication. Washington, DC: National Academy Press.

NCI (National Cancer Institute). 1997a. Estimated Exposures and Thyroid Doses Received by the American People from Iodine-131 in Fallout Following Nevada Atmospheric Nuclear Bomb Tests: Appendices to the Report from the National Cancer Institute. NIH Pub. No. 97-4264. Washington, DC: U.S. Department of Health and Human Services.

NCI (National Cancer Institute). 1997b. SEER Cancer Statistics Review. NIH Pub. No. 97-2789. Bethesda, MD: U.S. Department of Health and Human Services.

NCRP (National Council on Radiation Protection and Measurements). 1985. Induction of Thyroid Cancer by Ionizing Radiation. Bethesda, MD: NCRP.

Nease RF, Brooks WB. 1995. Patient desire for information and decision making in health care decisions: The autonomy preference index and the health opinion survey. Journal of General Internal Medicine 10:593-600.

Ng YC, Anspaugh LR, Cederwall RT. 1990. ORERP internal dose estimates for individuals. Health Physics 59:693-713.

NIH (National Institutes of Health). 1985. Report of the National Institutes of Health Ad Hoc Working Group to Develop Radioepidemiological Tables. NIH Publication 85-2748. Washington, DC: Government Printing Office.

NIH (National Institutes of Health). 1997. NCI releases study results of radioactive fallout from nuclear tests. CancerWEB. Available at: http://www.graylab.ac.ukcancernet/400206.html.

Nikiforov Y, Gnepp DR. 1994. Pediatric thyroid cancer after the Chernobyl disaster: Pathomorphologic study of 84 cases (1991-1992) from the Republic of Belarus. Cancer 74:748-766.

Nikiforov YE, Rowland JM, Bove KE, Monforte-Monoz H, Fagin JA. 1997. Distinct pattern of ret oncogene rearrangements in morphological variants of radiation-induced and sporadic thyroid papillary carcinomas in children. Cancer Research 57:1690-1994.

Nikiforova NV, Nedozhdy AV, Semushina SV, Krivyakova EV, Moiseyenko MY, Sivachenko TP, Elagin VV, Avramenko AI, Panasyuk GD, Derzhtskaya NK, Cot VA, Rafeyenko SM, Koulikova NV, Kroupnik TA, Steputin LA, Korobkova LP, Kolosvetova TY, Belova EA, Averichev AA, Saiko AS, Vaniliuk VV. 1996. Findings of the Chernobyl Sasakawa health and medical cooperation project: Goiter and iodine around Chernobyl. In: Yamashita S, Shibata Y (eds.), Chernobyl: A Decade. Proceedings of the Fifth Chernobyl Sasakawa Medical Cooperation Symposium, Kiev, Ukraine, 14-15 October 1986. Amsterdam, The Netherlands: Elsevier.

O'Connor AM. 1989. Effects of framing and level of probability on patients' preferences for cancer chemotherapy. Journal of Clinical Epidemiology 42:119-126.

O'Rahilly R, Muller F. 1992. Human Embryology and Teratology. New York: Wiley-Liss.

OECD (Organization for Economic Co-operation and Development). 1996. Chernobyl: Ten years on: Radiological and health impact. Paris: OECD/NEA.

Pacini F, Vorontsova T, Demidchik EP, Molinaro E, Agate L, Romei C, Shavrova E, Cherstvoy ED, Ivashkevitch Y, Kuchinskaya E, Schlumberger M, Ronga G, Filesi M, Pinchera A. 1997. Post-Chernobyl thyroid carcinoma in Belarus children and adolescents: Comparison with naturally occurring thyroid carcinoma in Italy and France. Journal of Clinical Endocrinology and Metabolism 82:3563-3569.

Papanicolaou Society of Cytopathology Task Force on Standards of Practice. 1996. Guidelines of the Papanicolaou Society of Cytopathology for the examination of fine-needle aspiration specimens from thyroid nodules. Modern Pathology 9(6):710.

Pauker SG, Kassirer JP. 1987. Decision analysis. New England Journal of Medicine 316(5):250-258.

Pauker SG, Kassirer JP. 1997. Contentious screening decisions: Does the choice matter? New England Journal of Medicine 336(17):1243-1244.

Pauker S, Kopelman RI. 1992. Interpreting hoofbeats: Can Bayes help clear the haze? New England Journal of Medicine 327(14):1009-1013.

Perloff RM. 1993. Third-person effect research 1983-1993: A review and synthesis. International Journal of Public Opinion Research 5:167-184.

Peters E, Slovic P. 1991. The role of affect and worldviews as orienting dispositions in the perception and acceptance of nuclear power. Journal of Applied Social Psychology 26(16):1427-1453.

Peters RG, Covello VT, McCallum DB. 1997. The determinants of trust and credibilty in environmental risk communication: An empirical study. Risk Analysis 17:43-54.

PHS (Public Health Service). 1963a. Consumption of Selected Food Items in U.S. Households—July 1962. Radiological Health Data 4. Bureau of the Census and Division of Radiological Health.

PHS (Public Health Service). 1963b. National Food Consumption Survey—Fresh Whole Milk Consumption in the United States—July 1962. Radiological Health Data 4. Bureau of the Census and Division of Radiological Health.

Pierce DA, Shimizu Y, Preston DL, Vaeth M, Mabuchi K. 1996. Studies of the mortality of atomic bomb survivors. Report 12, Part I. Cancer: 1950-1990. Radiation Research 146:1-27.

Piromalli D, Martelli G. 1992. The role of fine-needle aspiration in the diagnosis of thyroid nodules: Analysis of 795 consecutive cases. Journal of Surgical Oncology 50(4):247-250.

Pottern LM, Kaplan M, Larsen P, et al. 1990. Thyroid nodularity after childhood irradiation for lymphoid hyperplasia: A comparison of questionnaire and clinical findings. Journal of Epidemiology 43:449-460.

Ransohoff DF, Harris RP. 1997. Lessons from the mammography screening controversy: can we improve the debate? Annals of Internal Medicine 127(11):1029-1034.

Reeves B, Hawkins R. 1986. Effects of Mass Communication. Chicago: Science Research Associates.

Robbins J. 1994. Treatment of Thyroid Cancer in Childhood. Proceedings of a Workshop. DOE/EH-0426. Bethesda, MD: National Institutes of Health.

Robbins J, Adams W. 1989. Radiation effects in the Marshall Islands. In: Nagataki S (eds.), Radiation and the Thyroid. Amsterdam: Excerpta Medica. pp. 11-24.

Rogers EM, Storey JD. Communication campaigns. Berger CR, Chafee SH. 1987. In: Handbook of Communication Science. Newbury Park: Sage. pp. 817-846.

Ron E. 1996. Cancer risk following radioactive iodine-131 exposures in medicine. In: Proceedings of the thirty-second annual meeting of the National Council on Radiation Protection and Measurements. Arlington, VA: NCRP.

Ron E, Kleinerman R, Boice J, LiVolsi V, Flannery J, Fraumeni J. 1987. A population-based case-control study of thyroid cancer. Journal of the National Cancer Institute 79:1-12.

Ron E, Lubin JH, Shore RE, Mabuchi K, Modan B, Pottern LM, Schneider AB, Tucker MA, Boice Jr JD. 1995. Thyroid cancer after exposure to external radiation: A pooled analysis of seven studies. Radiation Research 141:259-277.

Ron E, Modan B, Preston D, Alfandary E, Stovall M, Boice J. 1989. Thyroid neoplasia following low-dose radiation in childhood. Radiation Research 120:516-531.

Rowan KE. 1990. Cognitive correlates of explanatory writing skill. Written Communication 7:316-341.

Rowan KE. 1994. The technical and democratic approaches to risk situations: Their appeal, limitations, and rhetorical alternative. Argumentation 8:391-409.

Rupp EM. 1980. Age dependent values of dietary intake for assessing human exposures to environmental pollutants. Health Physics 39:115-163.

Russell LB. 1994. Educated Guesses: Making Policy About Medical Screening Tests. Berkeley, CA: University of California Press.

Sanderson DCW, Allyson JD, Taylor AN, Rianin SN, Murphy S. 1994. An Airborne Gamma-Ray Survey of Parts of SW Scotland in February 1993. East Kilbryde, Scotland: Scottish University Research and Reactor Centre.

Sandman PM, Weinstein ND, Miller P. 1994. High risk or low: How location on a "risk ladder" affects perceived risk. Risk Analysis 14(1):35-45.

Santoro M, Chiappetta G, Cerrato A, Salvatore D, Zhang L, Manzo G, Picone A, Portella G, Santelli G, Vecchio G, Fusco A. 1996. Development of thyroid papillary carcinomas secondary to tissue-specific expression of the RET/PTC1 oncogene in transgenic mice. Oncogene 12:1821-1826.

Santoro M, Dathan NA, Berlingieri MT, Bongarzone I, Paulin C, Grieco M, Pierotti MA, Vecchio G, Fusco A. 1994. Molecular characterization of RET/PTC3: A novel rearranged version of the RET proto-oncogene in a human thyroid papillary carcinoma. Oncogene 9:509-516.

Schlumberger M, De Vathaire F, Travagli JP, Vassal G, Lemerle J, Parmentier C, Tubiana M. 1987. Differentiated thyroid carcinoma in childhood: Long term follow-up of 72 patients. Journal of Clinical Endocrinology and Metabolism 65:1088-1094.

Schlumberger MJ. 1998. Papillary and follicular thyroid carcinoma. New England Journal of Medicine 338(5):297-306.

Schneider AB, Bekerman C, Leland J, et al. 1997. Thyroid nodules in the follow-up of irradiated individuals: Comparison of thyroid ultrasound with scanning and palpation. Journal of Clinical Endocrinology and Metabolism 82:4020-4027.

Schneider AB, Ron E, Lubin J, Stovall M, Gierlowski TC. 1993. Dose-response relationships for radiation-induced thyroid cancer and thyroid nodules: Evidence for the prolonged effects of radiation on the thyroid. Journal of Clinical Endocrinology and Metabolism 77(2):362-369.

Schneider AB, Shore-Freedman E, Ryo UY, Bekersan C. 1985. Radiation-induced tumors of the head and neck following childhood irradiation: Prospective studies. Medicine 82:4020-4027.

Schooler C, Chaffee SH, Flora JA, Roser C. 1998. Health campaign channels: Tradeoffs among reach, specificity, and impact. Human Communication Research 24:410-432.

Schwartz L, Woloshin S. 1998. Communicating with the public about health risks. In: Workshop on Guidelines for Thyroid Cancer Screening. Washington, DC: Institute of Medicine.

Schwartz LM, Woloshin S, Black WC, Welch HG. 1997. The role of numeracy in understanding the benefit of screening mammography. Annals of Internal Medicine 127:966-972.

SEER (Surveillance, Epidemiology, and End Results). 1998. Available at: http://www.seer.ims.nci.nih.gov/.

Shaha AR, Loree TR, Shah JP. 1995. Prognostic factors and risk group analysis in follicular carcinoma of the thyroid. Surgery 118:1131-1138.

Shimaoka K, Bakri K. 1982. Thyroid screening program: Follow-up evaluation. New York State Journal of Medicine 82(8):1184-1187.

Shore RE. 1992. Issues and epidemiological evidence regarding radiation-induced thyroid cancer. Radiation Research 131:98-111.

Shore RE, Hildreth N, Dvoretsky P, Andresen E, Moseson M, Pasternack B. 1993. Thyroid cancer among persons given X-ray treatment in infancy for an enlarged thymus gland. American Journal of Epidemiology 137:1068-1080.

Simon SL. 1990. An analysis of vegetation interception data pertaining to close-in weapons test fallout. Health Physics 59:619-626.

Simon SL, Lloyd RD, Till JE, Hawthorne HA, Gren DC, Rallison ML, Stevens W. 1990. Development of a method to estimate thyroid dose from fallout radioiodine in a cohort study. Health Physics 59:669-691.

Simon SL, Robison WL. 1997. A compilation of nuclear weapons test detonation data for U.S. Pacific Ocean tests. Health Physics 73:258-264.

Slovic P. 1987. Perceptions of risk. Science 236:280-285.

Slovic P, Flynn JH, Layman M. 1991. Perceived risk, trust, and the politics of nuclear waste. Science 2554:1603-1607.

Slovic P, Krimsky S, Golding D. 1992. Perception of risk: Reflections on the psychometric paradigm. In: Social Theories of Risk. Westport, CT: Prager. pp. 117-152.

Smanik PA, Liu Q, Furminger TL, Ryu K, Xing S, Mazzaferri EL, Jhiang SM. 1996. Cloning of the human sodium iodide symporter. Biochemical and Biophysical Research Communications 226:339-345.

Sox HC. 1998. Benefit and harm associated with screening for breast cancer. New England Journal of Medicine 337(16):1145-1146.

Stark DD, Clark OH, Moss AA. 1983. High-resolution in ultrasonography and computed tomography of thyroid lesions in patients with hyperparathyroidism. Surgery 94(6):863-868.

Stevens W, Till JE, Thomas DC, Lyon JL, Kerber RA, Preston-Martin S, Simon SL, Rallison ML, Lloyd RD. 1992. Assessment of Leukemia and Thyroid Disease in Relation to Fallout in Utah. Report of a Cohort Study of Thyroid Disease and Radioactive Fallout from the Nevada Test Site. Salt Lake City, UT: University of Utah.

Stevenson RL, Greene MT. 1980. A reconsideration of bias in the news. Journalism Quarterly 57:115-121.

Stsjazhko VA, Tsyb AF, et al. 1995. Childhood thyroid cancer since the accident at Chernobyl [letter][comments]. British Medical Journal 310(6982):801.

Swelstad J, Scanlon EF, Murphy ED, Garces R, Kahndekar JD. 1977. Thyroid disease following irradiation for benign conditions. Archives of Surgery 112:380-383.

Takahashi T, Trott KR, Fujimori K, Simon SL, Ohtomo H, Nakashima N, Takaya K, Kimura N, Satomi S, Schoemaker MJ. 1997. An investigation into the prevalence of thyroid disease on Kwajalein Atoll, Marshall Islands. Health Physics 73:199-213.

Taki S, Kakuda K., Kakuma K, Annen Y. 1997. Thyroid nodules: Evaluation with US-guided core biopsy with an automated biopsy gun. Radiology 202(3):874-877.

Tan GH, Gharib H. 1997. Thyroid incidentalomas: Management approaches to nonpalpable nodules discovered incidentally on thyroid imaging. Annals of Internal Medicine 126:226-231.

Tan GH, Gharib H, Reading CC. 1995. Solitary thyroid nodules. Comparison between palpation and ultrasonography. Archives of Internal Medicine 155(22):2418-2423.

Taubes G. 1997. NCI reserves one expert panel, sides with another. Science 276:27-28.

Taurog A. 1996. Hormone synthesis. In: Braverman LE, Utiger RD (eds.), The Thyroid. A Fundamental and Clinical Text. 7th ed. Philadelphia: Lippincott-Raven Publishers. pp. 47-50.

Tell R, Sjödin H, Lundell G, Lewin F, Lewensohn R. 1997. Hypothyroidism after external radiotherapy for head and neck cancer. International Journal of Radiation Oncology, Biology, Physics 39:303-308.

Thompson D, Mabuchi K, Ron E, Soda M, Tokunaga M, Ochikubo S, Sugimoto S, Ikeda T, Terasaki M, Izumi S, Preston D. 1994. Cancer incidence in atomic bomb survivors. Part II: Solid tumors, 1958-1987. Radiation Research 137:S17-S67.

Thompson JC Jr. 1966. Variability of fluid milk consumption and its relationship to radionuclide intake. Radiological Health Data Report 7:139-144.

Thompson RS. 1996. What have HMOs learned about clinical prevention services? An examination of the experience at Group Health Cooperative of Puget Sound. The Milbank Quarterly 74(4):469-509.

Thorvaldsson SE, Tulinius H, Bjornsson J, Bjarnason O. 1992. Latent thyroid carcinoma in Iceland at autopsy. Pathology, Research and Practice 188(6):747-750.

Tichenor PJ, Donohue GA, Olien CN. 1980. Community Conflict and the Press. Beverly Hills, CA: Sage.

Travagli JP, Schlumberger M, De Vathaire F, Francese C, Parmentier C. 1995. Differentiated thyroid carcinoma in childhood. Journal of Endocrinological Investigation 18:161-164.

Trott KR, Schoemaker MJ, Fujimori K, Takahashi K, Simon SL, Nakashima N, Watanabe M, Satomi S. 1998. Thyroid cancer and thyroid nodules in the people of the Marshall Islands potentially exposed to radioactive fallout from nuclear weapons tests (abstract). In: International Symposium on Radiation and Thyroid Cancer, St. John's College, Cambridge, UK, July 20-23, 1998. Cambridge, UK. Abstract No. 7.

Trumbo C. 1995. Longitudinal Modeling of Public Issues: An Application of the Agenda-Setting Process to the Issue of Global Warming. Journalism and Mass Communication Monographs, No. 152.

Tu JV, Naylor CD. (Document of the Steering Committee of the Provincial Adult Cardiac Care Network of Ontario). 1996. Coronary artery bypass mortality rates in Ontario. A Canadian approach to quality assurance in cardiac surgery. Circulation 94(10):2429-2433.

Tucker MA, Jones P, Boice J, Robison L, Stone B, Stovall M, Jenkin R, Lubin J, Baum E, Siegel S, Meadows A, Hoover R, Fraumeni J. 1991. Therapeutic radiation at a young age is linked to secondary thyroid cancer. Cancer Research 51:2885-2888.

Turkelson C, Mitchell M. 1998. Evaluation of published guidelines for thyroid cancer screening. In: Workshop on Guidelines for Thyroid Cancer Screening. Washington, DC: Institute of Medicine.

Tyler TR. 1980. Impact of directly and indirectly experienced events: The origin of crime-related judgments and behaviors. Journal of Personality and Social Psychology 39:13-28.

Tyler TR. 1984. Assessing the risk of crime victimization: The integration of personal vicitimization experience and socially transmitted information. Journal of Social Issues 40:27-38.

Tyler TR, Cook FL. 1984. The mass media and judgments of risk: Distinguishing impact on personal and societal levels. Journal of Personality and Social Psychology 47:693-708.

Udelsman R. 1996. Parathyroid imaging: The myth and the reality. Radiology 201(2):317-318.

UNSCEAR (United Nations Scientific Committee on the Effects of Atomic Radiation). 1993. Sources and Effects of Ionizing Radiation. New York: United Nations.

USDA (U.S. Department of Agriculture). 1955. Food Consumption of Households in the U.S. Household Food Consumption Survey Report No. 1. Washington, DC: U.S. Department of Agriculture.

USPSTF (U.S. Preventive Services Task Force). 1996. Guide to Clinical Preventive Services: Second Edition. Washington, DC: U.S. Department of Health and Human Services.

Viswanathan K, Gierlowski TC, Schneider AB. 1994. Childhood thyroid cancer, characteristics and long-term outcome in children irradiated for benign conditions of the head and neck. Archives of Pediatric and Adolescent Medicine 148:260-265.

Voillequé PG, Kahn B, Krieger HL, Montgomery DM, Keller JH, Weiss BH. 1981. Evaluation of the Air-Vegetation-Milk Pathway for 131-I at the Quad Cities Nuclear Power Station. NUREGCR-1699. Washington, DC: U.S. Nuclear Regulatory Commission.

Volpé R. 1991a. Autoimmune thyroiditis. In: Braverman LE, Utiger RD (eds.), Werner and Ingbar's, The Thyroid. Philadelphia: J.B. Lippincott Co. pp. 921-933.

Volpé R. 1991b. Grave's disease. In: Braverman LE, Utiger RD (eds.), Werner and Ingbar's, The Thyroid. Philadelphia: J.B. Lippincott Co. pp. 648-657.

Vykhovanets EV, Chernyshov VP, Sluvkin II, Antipkin YG, Vasyuk AN, Klimenko HF, Strauss KW. 1997. ^{131}I dose-dependent thyroid autoimmune disorders in children living around Chernobyl. Clinical Immunology and Immunopathology 84:251-259.

Wang C, Crapo LM. 1997. The epidemiology of thyroid disease and implications for screening. Endocrinology and Metabolism Clinics of North America 26(1):189-218.

Wartella E, Reeves B. 1985. Historical trends in research on children and the media: 1900-1960. J. Comm. 35:116-133.

Watters DA, Ahuja AT. 1992. Role of ultrasound in the management of thyroid nodules. American Journal of Surgery 164(6):654-657.

Weart S. 1988. Nuclear Fear: A History of Images. Cambridge, MA: Harvard University Press.

Weiss BH, Voillequé PG, Keller PG, Kahn JH, Krieger HL, Martin A, Phillips CR. 1975. Detailed Measurements of ^{131}I in Air, Vegetation and Milk Around Three Operating Reactor Sites. NUREG-75/021. Washington, DC: U.S. Nuclear Regulatory Commission.

Wennberg DE, Lucas FL, Birkmeyer JD, Bredenberg CE, Fisher ES. 1998. Variation in carotid endarterectomy mortality in the medicare population: Trial hospitals, volume and patient characteristics. JAMA 279(16):1278-1281.

Werner A, Modan B, Davidoff D. 1968. Doses to brain, skull and thyroid, following X-ray therapy for tinea capitis. Physics in Medicine and Biology 13:247-258.

Williams ED. 1991. Biologic effects of radiation on the thyroid. In: Braverman LE, Utiger RD (eds.), Werner and Ingbar's, The Thyroid. Philadelphia: J.B. Lippincott Co. pp. 421-436.

Williams ED, Becker D, et al. 1996. Effects on the thyroid in populations exposed to radiation as a result of the Chernobyl accident. In: One decade after Chernobyl, Proceedings of an International Conference. Vienna: International Atomic Energy Agency.

Williams ED, Tronko ND. 1996. Molecular, Cellular, and Biological Characterization of Childhood Thyroid Cancer. International Scientific Collaboration on Consequences of Chernobyl Accident. Luxembourg: EUR: European Commission.

Wong FL, Ron E, Gierlowski T, Schneider AB. 1996. Benign thyroid tumors: General risk factors and their effects on radiation risk estimation. American Journal of Epidemiology 144:728-733.

Woolf SH. 1997. Should we screen for prostate cancer? Men over 50 have a right to decide for themselves. British Medical Journal 314:989-990.

WorthlinWorldwide. 1996. New study shows Internet use skyrockets and will continue as 42 million Americans go on-line. Available at: http://www.wirthlin.com/whatsnew/prs/internet.html.

Wynne B. 1991. Knowledges in context. Science, Technology, and Human Values 16:111-121.

Yang Y, Nelson CB. 1986. An estimation of daily food usage factors for assessing radionuclide intakes in the U.S. population. Health Physics 50:245-257.

Yoshimoto Y, Ezaki H, Etoh R, Hiraoka T, Akiba S. 1995. Prevalence rate of thyroid diseases among autopsy cases of the atomic bomb survivors in Hiroshima, 1951-1985. Radiation Research 141:278-286.

Zaller JR. 1992. The Nature and Origin of Mass Opinion. Cambridge, England: Cambridge University Press.

Zimmerman D, Hay I, Bergstrahl E. 1994. Papillary thyroid carcinoma in children. In: Robbins J. (ed.), Treatment of Thyroid Cancer in Childhood. Springfield, VA: National Technical Information Service. pp. 3-10.

Glossary

ABCC. See Atomic Bomb Casualty Commission.

Atomic Bomb Casualty Commission (abbreviated ABCC). The agency of the Japanese Ministry of Health and Welfare and the U.S. National Academy of Sciences charged with the responsibility for the study of the survivors of the atomic bombings from the inception of the investigations in 1947 until 1975.

Becquerel (abbreviated Bq). The measure of activity proposed by the Systeme International and equal to one nuclear disintegration per second.

Bq. See Becquerel.

Ci. See Curie.

Curie (abbreviated Ci). The measure of activity used until the introduction of the Becquerel. One Curie is equal to 3.7×10^{10} nuclear disintegrations per second or to 3.7×10^{10} Becquerels.

DDREF. See dose and dose-rate effectiveness factor.

Dose and dose-rate effectiveness factor (abbreviated DDREF). A measure of the extent to which radiation-related damage accruing at a high dose rate is ameliorated when the dose rate is low. This value will presumably vary with the endpoint measured, but it is not known precisely for such endpoints as incidence of or death due to cancer. Experimental studies suggest a value between 5 and 20, but the epidemiological data derived from the studies of the atomic bomb surveyors imply a lower number, possibly 1 to 2.

EAR. See excess risk.

ERR. See excess risk.

Excess risk. Risk, whether it is relative or absolute, is often expressed in terms of the excess that it represents over expectations in the absence of exposure to a carcinogenic agent, radiation in this instance. *Excess relative risk (ERR)* is, therefore, merely the observed relative risk minus 1, the value expected in the absence of an effect of radiation. Similarly *excess absolute risk (EAR)* is the number of incident cases or deaths at a particular dose above the number that would be expected in the absence of a radiation-related increase.

Exposure. Technically, exposure is the amount of air ionized by radiant energy, specifically, the amount of electrical charge produced in 1 cc of air under condition of electron equilibrium. More commonly, and in the present context, it simply means the presence of an individual in a field of radiation.

Geometric mean (abbreviated GM). The geometric average of a series of n positive numbers; it is equal to the nth root of their product. As an illustration, the geometric mean of the numbers 2 and 8 is the square root of 16 (their product) or 4. The GM is approximately the median (50th percentile) of a distribution of numbers whose logarithms are normally distributed.

Geometric standard deviation (abbreviated GSD). The geometric standard deviation is the antilog of the standard deviation of the natural logarithms of a set of numbers. The GSD is usually used to describe the variation of a set of data that is positively skewed.

GM. See geometric mean.

Gray (abbreviated Gy). The SI unit of absorbed dose equal to 1 joule per kilogram, or 100 rad. The unit derives its name from the English biophysicist Louis Harold Gray.

GSD. See geometric standard deviation.

Gy. See Gray.

Kriging. Kriging is defined as the process of estimating the value of a spatially distributed variable from adjacent values while considering their interdependence.

Probability of causation. A number that expresses the probability that a given cancer, in a specific tissue has been caused by a previous exposure to a carcinogenic agent, such as ionizing radiation.

rad. The unit of absorbed dose used prior to the introduction of the Gray. One rad is equal to 100 ergs per gram or 0.01 Gy.

Radiation Effects Research Foundation. The research institution currently charged with the study of the health effects of exposure to the atomic bombing of Hiroshima and Nagasaki; it is the successor to the Atomic Bomb Casualty Commission and is jointly administered and funded by the governments of Japan and the United States.

Rainout. The washing out of radionuclide-containing particles by rain falling to the earth's surface.

RBE. See relative biological effectiveness.

Relative biological effectiveness (abbreviated RBE). The biological effectiveness of one form of radiation as compared with another to produce the same biological endpoint. It affords the means to combine doses when an individual is exposed to a variety of forms of radiation.

rem. The "dose equivalent" (H) and its unit, the rem (roentgen equivalent-man), were introduced to account for the different biologic effects of the same absorbed dose from different types of radiation; H is the product of D, Q, and N at a point of interest in tissue, where D is absorbed dose, Q is the quality factor, and N is the product of any other modifying factors (1 rem = 0.01 Sv).

Risk, absolute (abbreviated AR). The excess number of deaths (or cases) above that "normally" expected in some population in the absence of exposure to ionizing radiation beyond that to which everyone is subjected because of the radiation emanating from the earth's crust or originating in outer space.

Risk, attributable. The percentage of deaths or cases ostensibly assignable to a specific cause, in this instance, ionizing radiation.

Risk, relative (abbreviated RR). The ratio of the risk in one population to that in another; for example, the ratio of the risk among individuals exposed to 2 Gy as contrasted with the background risk.

SEER. The Surveillance, Epidemiology, and End Results (SEER) Program was developed as a result of the National Cancer Act of 1971, which mandated the collection, analysis, and dissemination of all data useful in the prevention, diagnosis, and treatment of cancer. SEER is a continuing project of the National Cancer Institute to collect cancer data on a routine basis from designated population-based cancer registries in various areas of the country.

Sensitivity analysis. An analysis of the variation in the solution of a problem with variation in the values of the parameters involved. Basically, it is a method of determining how dependent a solution may be on the values assigned to the parameters in a model by the investigator(s).

Sievert (abbreviated SV). A unit of dose equivalent to the dose in Gray times a quality factor times any other factors that may modify the dose. The name is derived from the Swedish physicist Rolf M. Sievert.

SIR. Standardized incidence rate is the incidence (new cases in a time interval) rate adjusted to a population of standardized age distribution. It is usually equal to the ratio of the number of observed cases to the number of expected cases.

SMR. Standardized mortality rate is the mortality (death) rate adjusted to a population of a standardized age distribution. It is usually equal to the ratio of the number of observed deaths to the number of expected deaths.

Sv. See Sievert.

Uncertainty analysis. Models are merely approximations of real systems, and as such their predictions are inherently uncertain. Uncertainty analysis seeks to quantify the error inherent in each step in the modeling process and to propagate these errors through the entire process in order to estimate the overall possible error (or uncertainty).

Appendix A

Study Activities

To undertake this study, the IOM and the NRC established two committees (see rosters at the front of this report and the Committee Biographies). In its request to the IOM and the NRC, the NCI asked that the work of the Institute of Medicine and the National Research Council be conducted in public to the fullest extent possible. Consistent with the policies of the National Academy of Sciences, the committees conducted their fact-finding activities in public meetings and met in closed session only to consider findings and recommendations. The committees sought to hold as many open sessions as was consistent with its resources, charge, and Academy policy. At their first meetings in December 1997, both committees examined their composition to make certain that necessary expertise and perspectives were represented, and no significant conflicts of interest or bias existed. Those members of the committee who could not be present participated through a conference call.

The IOM Committee on Guidelines for Thyroid Cancer Screening after Exposure to Radioactive Iodine Fallout was appointed to focus on clinical and public health issues and policies. It included experts in preventive services, thyroid cancer diagnosis and management, medical decision-making, practice guideline development and implementation, and public health policy. The IOM committee met three times—in December 1997, March 1998, and April 1998. The first meeting on December 20 was a three-hour, closed organizing session. For the second meeting, the committee organized a March 17-18 workshop on thyroid cancer screening in which the NRC committee members also participated (see the end of this appendix for the agenda). At a closed one-day meeting on April 11, the committee reached final agreement on its conclusions and recommendations.

The NRC Committee on Exposure of the American People to I-131 from the Nevada Atomic Bomb Tests included people with expertise in thyroid disease, epidemiology, risk assessment, radiobiology, dose reconstruction, health physics, public health, risk communication, clinical practice, and medical ethics. The majority of this group, which focused on the first five tasks listed above, were already serving on the NRC Committee on an Assessment of CDC Radiation Studies and had accumulated considerable experience in dose reconstruction in the course of its work for the Centers for Disease Control and Prevention (CDC). Four individuals served on both committees.

The NRC committee met five times—in December 1997, January 1998, February 1998 (two meetings), and March 1998. In an open session December 19, the NRC committee and observers received some six hours of detailed briefings on the National Cancer Institute Report by the investigators. This began with an Introduction and Overview of the Study by Dr. Bruce Wachholz, and was followed by presentations on the "Estimation of the Activities of I-131 Deposited on the Ground" by Drs. Harold Beck and Lester Machta, on the "Transfer of I-131 from Deposition on the Ground to Fresh Cows' Milk" by Mr. Paul Voillequé and Drs. Mona Dreicer and André Bouville, on "Milk Production, Utilization, Distribution and Consumption" by Dr. Mona Dreicer, on "Dose Conversion Factors" by Dr. Jacob Robbins, on "Dose Reconstruction Methodology" by Dr. André Bouville, and finally on the "Estimated Thyroid Doses" by Drs. Lester Machta and Bruce Wachholz. These presentations were followed by a brief description of the methods employed to assess the risks of thyroid cancer resulting from the estimates of thyroid dose by Dr. Charles Land (see Appendix B). It should be noted that the estimates of the lifetime risk of thyroid cancer and the possible number of individuals so affected are not included in the two volume report but were provided to the committee in the form of a memorandum addressed to Dr. Richard Klausner, the Director of the National Cancer Institute, and subsequently a revised version as a communication to the committee. The questions raised by the committee centered on the reconstruction of the doses, and in particular, the use of kriging to estimate county values where direct measurements were not available, on the methods used to determine the uncertainties in the estimates of dose, and on the assumptions inherent in the estimates of the lifetime risk of thyroid cancer.

This briefing was followed by statements from Drs. Lynn Anspaugh, Donald Myers, Roy Shore, and F. Owen Hoffman, who performed supporting experimental research at Oak Ridge National Laboratory during the mid 1980s. Given the idiosyncratic nature of dose reconstructions, these four consultants to the committee were asked to present their views on the dosimetry and dose reconstruction used in the report, the risk factors developed by the investigators and epidemiological considerations for thyroid disease, and the appropriateness of the use of spatial interpolation and kriging to assign putative exposures where no direct measurements were available. Parenthetically, kriging is an algorithm for estimating the value of a spatially distributed variable from adjacent values while considering their interdependence. These presentations were followed by statements from the public including a prepared statement from Mr. E. Cooper Brown of the law firm of Cummins and Brown on the need for openness of all committee sessions.

On the second day of the NRC committee meeting, Dr. F. Owen Hoffman identified several issues relevant to the committee's charge that needed further discussion, and members of the IOM Committee on Thyroid Screening Related to I-131 Exposure, specifically Drs. Robert Lawrence and Steven Pauker, and co-study director Marilyn Field, discussed the approach being taken by the IOM in consideration of clinical and public health policies related to iodine-131 exposure and the risk of thyroid cancer. The NRC committee then grappled with the proposed structure of its report and the designation of writing assignments.

The second meeting of the NRC committee was held at the National Academy of Sciences' Beckman Center in Irvine, California on January 16-17, 1998. There the committee continued discussion of the charge, the structure of the report, and fact-finding. The first day of this meeting was in open session, and included three conference calls. The first of these involved Dr. Paul Gilman of the National Research Council's Commission on Life Sciences and dealt with the Academy's position and its legal obligations with respect to the issue of the openness of the committee's deliberations. The second was to Dr. Carl A. Gogolak of the Environmental Measurements Laboratory of the Department of Energy from whom the committee sought more information on how the kriging had been applied. Finally, the third was to Drs. Elaine Ron of the Radiation Epidemiology Branch of the National Cancer Institute and Jay A. Lubin of the Biostatistics Branch of the same institute who provided further information on the study they published in 1995 utilizing the data from some seven studies of the health effects of exposure to I-131. Their help was requested in further analyses of these data. This day's activities ended with presentations from three members of the public: Mr. James P. Thomas of Short, Cressman & Burgess, Attorneys-at-Law, in Seattle, Mr. Fred Allingham, Administrative Director of the National Association of Radiation Survivors, and Mrs. Kymberlee Burnell, a psychotherapist who spoke on her own behalf.

The second day focused on continued discussion of the structure of the report, and how best to integrate the work of the IOM committee investigating clinical and public health policies and the results to emerge from the Workshop on Thyroid Cancer Screening and Health Implications of Exposure to Radioactive Iodine Fallout to be held in Washington, D.C. on March 17-19, 1998. The meeting ended after agreement was reached on the dates of the next three committee meetings and further writing assignments were made.

The third meeting of the NRC committee occurred on February 10th and 11th, 1998, at the National Academy of Sciences in Washington. The first of these two days involved closed discussion of the report, and identification of areas where further fact-finding was indicated. The second day was open and attended by the committee's consultants as well as representatives of the public and other federal agencies whose activities impinge on the charge before the committee. Presentations were made by Dr. John Bagby, Mr. Tim Connor, and Ms. Trisha Pritikin, representing the Advisory Committee on Energy Related Epidemiological Research (ACERER) of CDC's National Center on Environmental Health (NCEH) and the National Institute of Occupational Safety and Health (NIOSH), by Dr. Robert Spengler of the Division of Health Studies of the Agency for Toxic Substances and Disease Registry (ATSDR), by Dr. Kenneth Kopecky of the Fred Hutchinson Cancer Research Center and one of the principal investigators in the Hanford Thyroid Disease Study, by Mr. Seth Tuler of the Center for Technology, Environment and Development (CENTED) of Clark University, and by Mr. James Thomas of Short, Cressman and Burgess, Attorneys-at-Law, in Seattle.

The fourth meeting of the NRC committee occurred on February 27th and 28th in Las Vegas, Nevada. On the first of these days the committee continued to develop its report and to explore alternative ways and language to respond to its charge. The second day of the meeting was open to the public and centered on discussion of the ways in which the risks and the public health consequences could best be communicated to the public. Some 14 members of the public or representatives of concerned groups were present and most participated in the discussion. To maximize interaction, the members of the committee discussed the charge before them, the nature of the National Research Council and its committee activities, and the time constraints under which the committee is operating. To further public understanding of the findings of the National Cancer Institute, these findings and those of the ORERP Study as well as the dose reconstruction methods they employed were contrasted by Dr. Lynn Anspaugh. In addition, on behalf of the committee, Drs. Keith Baverstock, F. Owen Hoffman, and Henry Royal presented a series of alternative ways in which risk might be presented for public reaction. Responses to these alternatives were made by Ms. Trisha Pritikin of ACERER, Mr. Richard A. Nielsen, Executive Director of Citizen Alert, and others.

As a possible paradigm for the NRC committee's recommendations to the Secretary of the Department of Health and Human Services, Ms. Bea Kelleigh, the Executive Director, Cedar River Associates, under contract with the Hanford Health Information Network to develop public information, described at some length the steps that have been and are being taken by the Hanford Health Information Network to inform the public of the findings of the studies associated with the Hanford Nuclear Facility and their public health implications.

On March 17, the committee met briefly in closed session in advance of the workshop organized by the IOM committee considering the clinical and public health policy implications of iodine-131 exposure and thyroid cancer risk. The workshop agenda and participants list is provided below.

WORKSHOP ON THYROID CANCER SCREENING AND HEALTH IMPLICATIONS OF EXPOSURE TO RADIOACTIVE IODINE FALLOUT

March 17-18, 1998 • Washington D.C.

AGENDA

TUESDAY, MARCH 17 LECTURE ROOM, 2101 Constitution Avenue NW

1:00 **Welcome, Introductions, and Workshop Overview**
Robert Lawrence, M.D., IOM Workshop Chair; Johns Hopkins University School of Public Health
Marilyn Field, Ph.D., Co-Study Director, IOM
Charles Land, Ph.D., Project Officer, National Cancer Institute
James Smith, Ph.D., Centers for Disease Control and Prevention

1:20 **Review of I-131 Exposure Analysis and Estimates**
William Schull, Ph.D., Chair, NRC Committee on I-131 Exposure
University of Texas, Houston School of Public Health

1:40 **Overview of Thyroid Cancer and Other Thyroid Disease**
Natural History, Management, and Prevalence
Ernest Mazzaferri, M.D., IOM Committee Member
Ohio State University Medical Center
Description of Screening and Diagnosis Options
William Charboneau, M.D., IOM Committee Member
Mayo Clinic
Virginia LiVolsi, M.D., IOM Committee Member
Hospital of the University of Pennsylvania

2:20 Break

2:40 **Developing and Implementing Clinical Practice Guidelines for Cancer Screening**
Harold Sox, M.D., Chair, U.S. Preventive Services Task Force
Dartmouth-Hitchcock Medical Center
Lisa Schwartz, M.D. and Steven Woloshin, M.D., Dartmouth Medical School; Veterans Affairs Outcomes Group White River VAMC
David Ransohoff, M.D., University of North Carolina, Chapel Hill
Robert Thompson, M.D., Group Health Cooperative of Puget Sound
General discussion

4:40 **Heuristic Decision Model for Thyroid Cancer**
Stephen Pauker, M.D., IOM Committee Member
New England Medical Center
General discussion

WEDNESDAY, MARCH 18 MEMBERS ROOM, 2101 Constitution Avenue NW

9:00 **Review of the Literature on Thyroid Cancer Screening**
Mark Helfand, M.D. and Karen Eden, Ph.D.
Oregon Health Sciences University
Recommendations of Others
Charles Turkleson, Ph.D. and Matthew Mitchell, Ph.D., ECRI

9:45 **Discussants**:
Arthur Schneider, M.D., University of Illinois, Chicago
Barbara Carroll, M.D., Duke University
Roy Shore, M.D., New York University
Paul Ladenson, M.D., Johns Hopkins University
Robert Spengler, Sc.D. and Richard Reingans, Ph.D., Agency for Toxic Substances and Disease Registry, CDC
General discussion

12:00 Lunch

1:15 **Continued discussion**

2:00 **Public Health and Consumer Perspectives**
Kristine Gebbie, M.D., Former Commissioner of Health, State of Washington; Columbia University
Morton Rabinowitz, Ph.D., DuPont
H. Jack Geiger, M.D., City University of New York Medical School
Richard Schultz, M.S., Idaho Department of Health and Welfare

3:30 Break

3:45 **Public Comment Period**

5:45 **Adjourn**

WORKSHOP PARTICIPANTS LIST

Keith F. Baverstock, Ph.D.
Head of the Radiation Protection
 Division
World Health Organization

David V. Becker, M.D.
Professor of Radiology and Nuclear
 Medicine
New York Hospital

Clyde Behney
Deputy Executive Officer
Institute of Medicine/National Academy of Sciences

Stephen A. Benjamin, Ph.D.
Professor of Pathology, Radiological
 Health Sciences and Environmental
 Health
Colorado State University

Catherine Borbas, Ph.D.
Executive Director
Healthcare Education and Research
 Foundation, Inc.

A. Bertrand Brill, M.D., Ph.D.
Professor of Radiology/Physics,
 Biomedical Engineering
Vanderbilt University Medical Center

J. William Charboneau, M.D.
Professor of Radiology
Mayo Clinic and Mayo Medical
  School

Evan B. Douple, Ph.D.
Director, Board on Radiation Effects
  Research
National Research Council

Sharon Dunwoody, Ph.D.
Evjue-Bascom Professor of Journal-
  ism and Mass Communication,
  Head of Academic Programs,
  Institute for Environmental Studies
University of Wisconsin-Madison

Karen Beekman Eden, Ph.D.
Assistant Professor
Oregon Health Sciences University

Marilyn J. Field, Ph.D.
Co-study Director, I-131 report
Deputy Director of Health Care
  Services
Institute of Medicine

Kristine M. Gebbie, Dr.P.H., R.N.
Assistant Professor, Center for Health
  Policy and Health Services Re-
  search
Columbia University School of
  Nursing

H. Jack Geiger, M.D.
Arthur C. Logan Professor Emeritus,
  Dept. of Community Health and
  Social Medicine
City University of New York Medical
  School

Paul Gilman, Ph.D.
Executive Director
Commission on Life Sciences
National Research Council/National
  Academy of Sciences

Peter G. Groer, Ph.D.
Associate Professor
University of Tennessee
Department of Nuclear Engineering

Michael F. Hartshorne, M.D.
Professor and Vice Chairman of
  Radiology
University of New Mexico and Joint
  Imaging Service VAMC Albuquer-
  que

Karen Hein, Ph.D.
Executive Officer
Institute of Medicine/National Acad-
  emy of Sciences

Mark Helfand, M.D.
Assistant Professor, Division of
  Medical Informatics and Outcomes
  Research
Oregon Health Sciences University

Allan Korn, M.D.
Vice President and Medical Director
Blue Cross and Blue Shield of Illinois

Paul Ladenson, M.D.
Director of the Division of Endocri-
  nology and Metabolism
Johns Hopkins University School of
  Medicine

Susan Lederer, Ph.D.
Associate Professor of Humanities
Pennsylvania State University

Laura Leonard
Project Coordinator, Hanford Health
  Information Network
Oregon Health Division

Virginia A. LiVolsi, M.D.
Vice Chair for Anatomic Pathology
Hospital of the University of Pennsyl-
  vania

James E. Martin, Ph.D.
Associate Professor of Radiological
  Health
University of Michigan

Ernest L. Mazzaferri, M.D.
Professor and Chair, Department of
  Internal Medicine
Ohio State University

Kathryn Merriam, Ph.D.
Synthesis, Inc.

Matthew D. Mitchell, Ph.D.
Technology Analyst
ECRI, Inc.

Christopher B. Nelson, B.S.
Environmental Engineer
Environmental Protection Agency

Stephen G. Pauker, M.D.
Vice Chair for Clinical Affairs
New England Medical Center
Department of Medicine

Morton Rabinowitz, M.D.
Media, Pennsylvania

David Ransohoff, M.D.
Professor
University of North Carolina, Chapel
  Hill

Richard Reingans, Ph.D.
Health Economist
Agency for Toxic Substances and
  Disease Registry

Jacob Robbins, M.D.
Senior Scientist Emeritus, National
  Institute of Diabetes and Digestive
  and Kidney Diseases (NIDDK)
National Institutes of Health

Henry D. Royal, M.D.
Associate Director, Division of
  Nuclear Medicine
Mallinckrodt Institute of Radiology

Arthur B. Schneider, M.D., Ph.D.
Section Chief, Department of Endo-
  crinology-M/C 640
University of Illinois at Chicago

William J. Schull, Ph.D.
Chairman
Center for Demographic and Popula-
  tion Genetics
School of Public Health
University of Texas

Richard H. Schultz, M.S.
Administrator
State of Idaho
State Division of Health
Department of Health and Welfare

Lisa Schwartz, M.D.
Assistant Professor of Medicine
Dartmouth Medical School

Roy Shore, Ph.D.
Professor, Environmental Medicine
New York University Medical Center
Laboratory of Epidemiology and
  Biostatistics

Steven L. Simon, Ph.D.
Co-study Director, I-131 report
Senior Staff Officer
National Research Council

James Smith, Ph.D.
Chief of Radiation Studies Branch,
  Division of Environmental Hazards
  and Health Effects
Centers for Disease Control and
  Prevention

Harold C. Sox, M.D.
Professor and Chair
Dartmouth-Hitchcock Medical Center
One Medical Center Drive

Robert F. Spengler, Sc.D.
Division of Health Studies
Agency for Toxic Substances and
  Disease Registry

Michael Stoto, Ph.D.
Senior Program Officer, Division of
Health Promotion and Disease
Prevention
Institute of Medicine/National Acad-
emy of Sciences

Daniel O. Stram, Ph.D.
Associate Professor
University of Southern California
Department of Preventive Medicine

Robert G. Thomas, PhD
Kallispell, MT

Robert S. Thompson, M.D.
Director, Department of Preventive
Care
Group Health Cooperative of Puget
Sound

Charles M. Turkelson, Ph.D.
Chief Research Analyst
ECRI

R. Michael Tuttle, M.D.
Assistant Chief, Department of Clini-
cal Investigation; Chief, Clinical
Studies Service
Walter Reed Army Medical Center

Samuel A. Wells, M.D.
Professor of Surgery
Washington University School of
Medicine

Steven Woloshin, M.D.
Assistant Professor of Medicine
Dartmouth Medical School

Steven H. Woolf, M.D., M.P.H.
Professor, Department of Family
Practice
Virginia Commonwealth University
Medical College of Virginia

Appendix B

Copy of the Memorandum from Dr. Charles Land to Dr. Richard Klausner

DATE: September 23, 1997
TO: Dr. Richard Klausner, Director, NCI
FROM: Charles Land, Ph.D., Health Statistician, DCEG/EBP/REB
THROUGH: Director, DCEG
SUBJECT: Calculation of lifetime thyroid cancer risk for an average thyroid dose of 0.02 Gy from I-131 in fallout

My calculations of the thyroid cancer risk that might be associated with exposure to the American public to ^{131}I fallout from the Nevada Test Site resulted in an estimated range of 7,500 to 75,000 excess thyroid cancers during the lifetime of those exposed before 20 years of age. This range of estimates may be compared with about 400,000 expected, according to current SEER rates, among this segment of the US population. Thus, the estimated excess is between 2% and 19% of what might be expected in the absence of exposure. The calculations were based on a published, pooled analysis of thyroid cancer risk data from 5 cohort studies of populations exposed during childhood to medical x ray, or to gamma ray from the atomic bombings of Hiroshima and Nagasaki (Ron et al., 1995). They also incorporate various assumptions about the relative biological effectiveness (RBE) of ^{131}I compared to x ray or gamma ray. Significant excess risk was assumed to occur only following exposure before 20 years of age, in accordance with the epidemiological literature. A linear dose response was assumed, and the dose-specific excess relative risk, which was assumed to decrease sharply with increasing age at exposure, was also assumed to remain constant over the lifetime of the exposed population.

The calculations (see attached Excel spreadsheet)

Column 1 identifies the exposure ages considered. The first year of life was treated separately and older ages were grouped: 1-4, 5-9, 10-14, and 15-19. Exposure at ages older than 20 was ignored because there is little or no evidence of an excess cancer risk associated with exposure in adult life even to gamma and x-ray irradiation. Columns 2 and 3 give the estimated number of persons in the 1952 population of the US, by age and sex, as interpolated from 1950 and 1960 census numbers. The total number exposed at ages 0-19 also includes persons born in 1953, 1954, etc., but the entry into the population of newborn persons is largely compensated by the loss of persons reaching age 20 in the same years. With a linear dose-response model and lifetime excess risk, error introduced by acting as if the population 0-19 years of age in 1952 received all the dose that was actually received by those who were 0-19 years old during any part of the above-ground testing period is relatively unimportant.

Column 4 gives age-specific average thyroid doses in rad corresponding to the assumed average dose of 2 rad (0.02 Gy), based on information provided by André Bouville (this is why the first year of life was separated from the next four). As you know, thyroid doses to children are larger than those for adults because of smaller gland size, higher milk intake, and higher metabolism.

Column 5 gives the age-specific, linear dose-response coefficients for x ray and gamma ray, derived from Ron et al. (1995). Their overall coefficient for excess relative risk (ERR) at 1 rad was 0.077. They also did analyses suggesting that the ERR decreases by a factor of 2 for each successive 5-year interval of age at exposure, over the range 0-14 years of age. I derived the values in column 5 from the Ron et al. analysis, and extended the 2-fold reduction rule to 15-19 years at exposure. In each subsection, the age-specific coefficients have been multiplied by the specified RBE value.

Columns 6 and 7 are the estimated lifetime excess thyroid cancer rates for males and females, computed by multiplying the product of columns 4 and 5 by 0.25% for males and 0.64% for females, respectively; these percentages are the SEER (1973-1992) report's estimated lifetime thyroid cancer rates for men and women. The 1973-1994 SEER volume is now out, and gives 0.27% for males and 0.66% for females. Use of the new values would increase the total by about 4%.

Columns 8 and 9 were obtained by multiplying columns 2 and 3 by columns 6 and 7, respectively, and column 10 is the sum of columns 8 and 9. One implication of column 10 is that 75% of all the excess risk is estimated to result from exposure during the first 5 years of life.

The calculations are repeated for RBE values of 1.0, 0.66, 0.3, and 0.1.

Sources of uncertainty

NCRP report No. 80, "Induction of Thyroid Cancer by Ionizing Radiation," 1985, gave a range of 0.1 to 1.0 for the RBE of thyroid dose from ingested or inhaled I-131 compared to gamma ray or x ray, based on experimental studies. The report recommended 0.3 for radiation protection purposes, as the highest credible value. The NCRP report also stated that the RBE of ^{131}I relative to x ray may be lower at high doses and dose rates, and higher (nearer to x ray in effectiveness) at low dose and dose rates. Thus, Walinder (1972, summarized in the NCRP report) obtained an RBE of 0.1 using ^{131}I thyroid doses in the range 2200-11,000 rad whereas Lee et al. (1982) found near equivalence using dose groups at 80, 330 and 850 rad. Laird (1987) conducted parallel and combined analyses of 6 cohorts of children exposed to external radiation and one exposed to ^{131}I, and reevaluated experimental data from the large study of Lee et al. (1982) specifically designed to investigate the RBE of [131]I. Her RBE estimate was 0.66 with 95% confidence limits 0.14-3.15 (however, there is no support that I know of for an RBE greater than 1). The RBE value at low doses remains a contentious issue.

The range of estimates does not take into account statistical uncertainty about the Ron coefficients or statistical and subjective uncertainty about the estimated average dose. The Ron estimate of $ERR_{1Gy} = 7.7$ had 95% confidence limits 2.1-28.7, corresponding to a geometric standard deviation (GSD) of about 1.95. The average dose estimated by NCI, 2 rad, was assigned a GSD of 3, and therefore the product of that dose and the estimated ERR at 1 rad has a GSD of 3.6 (calculated as the exponential of the square root of the sum of squares of the natural logarithms of 1.95 and 3). Approximate 95% confidence limits for the number of excess cases are obtained by dividing and multiplying by 12.4 (= $3.6^{1.96}$). Thus, for example, ignoring all other possible sources of error, an estimate of 49,000 lifetime excess cases (corresponding to RBE = 0.66) would have confidence limits 4,000-608,000.

According to the model used for the estimates, ERR is constant over time following exposure, and about one third of the total excess lifetime risk among men in the exposed population, and about half among women, should already have taken place. It is possible, however, that the actual excess relative risk per unit dose may decline over time following exposure, most of which occurred over 40 years ago. Ron et al. found significant variation by time following exposure, but did not find a statistically significant trend. At the present time there are few data on radiation-related thyroid cancer risk 40 or more years following exposure during childhood, and therefore little basis for a discussion of the question.

References

Laird NM. Thyroid cancer risk from exposure to ionizing radiation: a case study in the comparative potency model. Risk Analysis 1987; 7: 299-309.

Lee W, Chiaccierini RP, Schlein B, Telles NC. Thyroid tumors following I-131 or localized x irradiation to the thyroid and the pituitary glands in rats. Radiation Research 1982; 92: 307-319.

NCRP Report No. 80. Induction of thyroid cancer by ionizing radiation. National Council on Radiation Protection and Measurements, Bethesda, 1985.

Ron E, Lubin JH, Shore RE, Mabuchi K, Modan B, Pottern L, Schneider AB, Tucker MA, Boice JD Jr. Thyroid cancer after exposure to external radiation: a pooled analysis of seven studies. Radiat Res 1995; 141: 259-77.

Walinder G. Late effects of irradiation on the thyroid gland of CBA mice. I. Irradiation of adult mice. Acta Radiol Ther Phys Biol 1972; 11: 433.

Age at exposure	1952 population count Males	Females	Estimated average thyroid dose (rad)	ERR at 1 rad (Ron et al. 1995)	Excess rate (lifetime) Males	Females	Lifetime excess thyroid cancers Males	Females	Total
				RBE = 1					
0	1,757,800	1,698,600	10.3	0.098	0.002524	0.00646	4,435.80	10,973.20	15,409.00
1-4	7,171,000	6,931,200	6.7	0.098	0.001642	0.004202	11,771.20	29,126.60	40,897.80
5-9	7,174,043	6,929,430	4.5	0.049	0.000551	0.001411	3,954.70	9,778.80	13,733.50
10-14	6,235,357	6,023,970	2.8	0.0245	0.000172	0.000439	1,069.50	2,644.80	3,714.30
15-19	5,916,664	5,715,114	1.8	0.01225	5.51e-05	0.000141	326.2	806.5	1,132.70
Total	28,255,864	27,298,314					21,557	53,330	74,887
				RBE = 0.66					
0	1,757,800	1,698,600	10.3	0.06468	0.001666	0.004264	2,927.60	7,242.30	10,170.00
1-4	7,171,000	6,931,200	6.7	0.06468	0.001083	0.002773	7,769.00	19,223.50	26,992.50
5-9	7,174,043	6,929,430	4.5	0.03234	0.000364	0.000931	2,610.10	6,454.00	9,064.10
10-14	6,236,357	6,023,970	2.8	0.01617	0.000113	0.00029	705.9	1,745.50	2,451.40
15-19	5,916,664	5,715,114	1.8	0.008085	3.64e-05	9.31e-05	215.3	532.3	747.6
Total	28,255,864	27,298,314					14,228	35,198	49,426
				RBE = 0.3					
0	1,757,800	1,698,600	10.3	0.0294	0.000757	0.001938	1,330.70	3,292.00	4,622.70
1-4	7,171,000	6,931,200	6.7	0.0294	0.000492	0.001261	3,531.40	8,738.00	12,269.30
5-9	7,174,043	6,929,430	4.5	0.0147	0.000165	0.000423	1,186.40	2,933.6	4,120.10
10-14	6,236,357	6,023,970	2.8	0.00735	5.15e-05	0.000132	320.9	793.4	1,114.30
15-19	5,915,664	5,715,114	1.8	0.003675	1.65e-05	4.23e-05	97.8	242	339.8
Total	28,255,864	27,298,314					6,467	15,999	22,466
				RBE = 0.1					
0	1,757,800	1,698,600	10.3	0.0098	0.000252	0.000646	443.6	1,097.30	1,540.90
1-4	7,171,000	6,931,200	6.7	0.0098	0.000164	0.00042	1,177.10	2,912.70	4,089.80
5-9	7,174,043	6,929,430	4.5	0.0049	5.51e-05	0.000141	395.5	977.9	1,373.40
10-14	6,236,357	6,023,970	2.8	0.00245	1.72e-05	4.39e-05	107	264.5	371.4
15-19	5,916,664	5,715,114	1.8	0.001225	5.51e-06	1.41e-05	32.6	80.7	113.3
Total	28,255,864	27,298,314					2,156	5,333	7,489

**Written amendment to 23 September memo, provided by Charles Land to
NAS committee on 19 December, 1997**

Calculation of the Estimated Lifetime Risk of Radiation-Related Thyroid Cancer in the U.S. Population from NTS Fallout

1. Thyroid cancer risk associated with gamma-ray and x-ray exposure, from studies of the Hiroshima-Nagasaki survivors and of various medically-exposed populations, is well quantified. Findings are summarized in a pooled analysis of seven studies (Ron et al., Radiation Research 1995; 141:259-277).

   • The evidence for a radiation-related risk is strong for childhood exposure, and weak or non-existent for adult exposure.

   • Dose-specific excess risk decreases with increasing age at exposure. At ages 5-9, it is about half that associated with exposure at ages 0-4, and at 10-14 it is about half that at 5-9.

   • For any given exposure age, excess risk appears to be proportional to thyroid dose (linear dose response).

   • Ron et al. estimated an excess relative risk (ERR) of 7.7 per Gy, or 0.077 per rad, for childhood exposure at ages younger than 15.

2. The average (case-weighted) exposure age in the pooled data was a little over $4^1/_2$ years. By linear interpolation between the midpoints of the first two intervals, and extension of the observed reduction in ERR with increasing age at exposure, the following age-specific coefficients were inferred:

Age at exposure	ERR at 1 rad
0-4	0.098
5-9	0.049
10-14	0.0245
≥ 20	negligible

3. Although there was evidence of variation radiation-related relative risk over time following exposure, there was no evidence of a trend. Accordingly, ERR was assumed to remain constant over the remainder of life.

4. Data on risk associated with thyroid exposure from ingested or inhaled ^{131}I suggest that there is a risk, but precise dose-response estimates are not available. Accordingly, it is reasonable to use the coefficients developed from data on x-ray and gamma-ray exposure, with an appropriate value for the relative biological effectiveness of ^{131}I compared to gamma rays or x rays.

   • NCRP report No. 80, "Induction of Thyroid Cancer by Ionizing Radiation," 1985, gave a range of 0.1 to 1.0 for the RBE, based on experimental studies. The report recommended 0.3 for radiation protection purposes, as the highest credible value. The NCRP report also stated that the RBE of

[131]I relative to x ray may be lower at high doses and dose rates, and higher (nearer to x ray in effectiveness) at low doses and dose rates.

- Thus, Walinder (1972, summarized in the NCRP report) obtained an RBE of 0.1 using ^{131}I thyroid doses in the range 2200-11,000 rad, whereas

- Lee et al. (1982) found near equivalence using dose groups at 80, 330, and 850 rad.

- Laird (1987) conducted parallel and combined analyses of 6 cohorts of children exposed to external radiation and one exposed to ^{131}I, and re-evaluated experimental data from the large study by Lee et al. (1982) specifically designed to investigate the RBE of [131]I. Her RBE estimate was 0.66 with 95% confidence limits 0.14-3.15 (however, there is no support that I know of for an RBE greater than 1).

- The RBE value at low doses remains a contentious issue.

- In the calculations for NCI, RBE values of 1, 0.66, 0.33, and 0.1 were assumed.

5. In addition to being more sensitive to the carcinogenic effects of ionizing radiation, the thyroid glands of children receive higher doses from ingested or inhaled ^{131}I than do the glands of adults, because of smaller gland size, higher intake of milk, and higher metabolism. Using conversion factors obtained from Dr. Bouville, the estimated average thyroid dose of 2 rad to the U.S. population from Nevada Test Site fallout was converted to the following values for children:

Exposure Age	Estimated Average Dose
<1	10.3
1-4	6.7
5-9	4.5
10-14	2.8
15-19	1.8

6. Lifetime cumulative thyroid cancer incidence rates of 0.25% for males and 0.64% for females, respectively, were assumed, based on the SEER report for 1973-1992. The 1973-1994 SEER volume is now out, and gives 0.27% for males and 0.66% for females. Use of the new values would increase the total by about 4%.

7. For simplicity of calculation, it was assumed that the U.S. population in 1952 received the total thyroid dose from NTS fallout in that year, instead of spread out over 12 years. This simplification was possible because, using a linear dose-response model, lifetime radiation-related thyroid cancer risk is proportional to summed collective dose, in person-rads, over exposure ages weighted by age-specific risk coefficient.

8. For each single year of age (column 1 in the spreadsheet), the sex-specific estimated numbers of lifetime excess thyroid cancer cases in the US due to NTS fallout (columns 8 and 9) were obtained as the product of:

    *   the number of male or female persons in the 1952 US population (columns 2 and 3)

    *   the age-specific estimated average cumulative thyroid dose over the entire period of above-ground testing (column 4)

    *   the age-specific linear dose-response coefficient (ERR at 1 rad) for x ray and gamma ray (column 5), times the assumed RBE for ^{131}I

    *   the cumulative lifetime thyroid cancer risk for men or women (0.25% or 0.64%), as appropriate.

9. The age and sex-specific totals were summed over sexes (column 10) and ages. The sums are given below columns 8-10 in each table.

10. Besides uncertainty about the RBE, there is also statistical uncertainty about the risk coefficients, and subjective and statistical uncertainty about the average doses used. The combined uncertainty is substantial. For example:

    *   95% confidence limits (2.1-28.7) for the Ron estimate of $ERR_{1Gy} = 7.7$ correspond approximately to a lognormal model geometric standard deviation (GSD) of about 1.95.

    *   The uncertainty of average dose estimated by the NCI, 2 rad, was stated to be between 1 and 4, i.e., a factor of 2 in each direction. This corresponds approximately to 95% confidence limits and thus to a GSD of about 1.4.

    *   Therefore, the product of that dose and the estimated ERR at 1 rad has a GSD of 2.1 (calculated as the exponential of the square root of the sum of squares of the natural logarithms of 1.95 and 1.4).

    *   Approximate 95% confidence limits for the number of excess cases can be obtained by dividing and multiplying by 4.3 (=$2.1^{1.96}$). Thus, for example, ignoring all other possible sources of error, an estimate of 49,000 lifetime excess cases (corresponding to a fixed RBE of 0.66, which here is assumed to be known without error) might be given with uncertainty 11,300-212,000.

Appendix C

Calculation of Collective Thyroid Dose to the U.S. Population from the Release of ^{131}I from the Nuclear Weapons Tests in Nevada

In an attempt to corroborate findings of the National Cancer Institute (see discussion in Chapter 2), a simple calculation is presented to estimate the collective dose to the thyroids of the U.S. population that occurred from the release of ^{131}I from the nuclear weapons tests carried out in Nevada during the 1950s and 1960s. The collective dose can be estimated in an approximate way by use of the following equation:

$$D = C \times S \times f_d \times \frac{1}{A} \times \alpha \times MPD \times f_m \times L \times N \times DCF \times \int_0^\infty e^{-\lambda t}\, dt, \qquad (1)$$

which has the solution of

$$D = C \times S \times f_d \times \frac{1}{A} \times \alpha \times MPD \times f_m \times L \times N \times DCF \times \frac{1}{\lambda}. \qquad (2)$$

In eqns (1) and (2) the parameters are defined as follows:

C	=	Units-conversion constant, 390,625 mi^{-2} m^2 μCi MCi^{-1};
S	=	Amount of ^{131}I released by the tests, MCi;
f_d	=	Fraction of the ^{131}I released that is deposited within the continental U.S.;
A	=	Area of the continental U.S., mi^2;
α	=	Mass-interception fraction for fallout retained on vegetation, m^2 kg^{-1};
MPD	=	Mass (dry) per day of pasture or green chop consumed by a cow, kg day^{-1};
f_m	=	Fraction of cow's intake that is secreted in milk, day L^{-1};
L	=	Consumption rate of milk, L day^{-1};

N = Number of persons in the U.S. in 1955;
DCF = Dose-conversion factor for the thyroid, rad mCi^{-1}; and
λ = Effective rate of loss of ^{131}I on pasture, day^{-1}.

The amount, S, of ^{131}I released by the tests at the Nevada Test Site was given in the NCI report from data derived by Hicks (1982, 1990). A reasonable estimate is 150 MCi, which may be considered as a geometric mean with a geometric standard deviation of 1.25. The latter value is based upon an estimate for a closely related parameter and is given by Ng et al. (1990).

The fraction, f_d, of the ^{131}I released by the tests that deposits within the continental United States is much more difficult to quantify. In a study to be published, Anspaugh and McArthur (1998) have determined that the amount of ^{137}Cs remaining in the NTS and the Phase I and II areas of the ORERP study (see Church et al. 1990) is equivalent to 10 percent of that originally released. This may be considered as a lower bound on the amount deposited within the continental U.S. An upper bound could be considered as 1.00. Then, assuming that these bounds are the 5th and 95th percentile of a lognormal distribution, the geometric mean is 0.32 with a geometric standard deviation of 2.0. This value is also reasonably consistent with the independent estimate made by Beck et al. (1990) that 25-30 percent of ^{137}Cs produced in the Nevada Tests was deposited within the continental U.S. The latter estimate was based upon drawing crude contours of deposition density as measured by gummed-film collectors and calculating that the fraction of ^{137}Cs deposited within those contours was 21 percent (the estimates ranged from 5 percent for Operations Ranger and Hardtack II to 42 percent for Operation Tumbler-Snapper). An additional amount of about 10 percent was assumed to be deposited with local fallout; thus, the estimate of 25-30 percent of total fallout deposited within the continental U.S.

The area, A, of the continental U.S. according to data in Funk & Wagnalls (1994) is 3,119,963 square miles.

The mass-interception fraction, α, of ^{131}I in fallout that is retained by pasture-type vegetation is taken to be 1.0 based on the article by Simon (1990). Further, the range is estimated to be from 0.5 to 2.0. This is approximately equivalent to a geometric mean of 1.0 with a geometric standard deviation of 1.52.

The dry mass consumption rate, MPD, for milk cows is based on the work by Koranda (1965), who reported an average value of 14 kg day^{-1}. It is assumed here that this value can be characterized by a geometric standard deviation of 1.5; this would correspond to a geometric mean value of 13 kg day^{-1}.

The classic value for the rate of secretion of ^{131}I in milk, f_m, is 0.005 day L^{-1} (Garner and Russell 1966). This value is assumed to be a geometric mean with a geometric standard deviation of 1.5.

The per capita consumption rate of milk, L, is taken from the NCI report. The value of 0.4 L day^{-1} is assumed to be a geometric mean with a geometric standard deviation of 1.5.

The number of persons in the continental U.S. in 1955, N, is interpolated between the 1950 and 1960 census values reported in Funk & Wagnalls (1994). This value is 165,000,000.

The population-weighted dose-conversion factor for ^{131}I-thyroid dose is about 2 rad mCi[-1]. This value is assumed to be a geometric mean with a geometric standard deviation of 1.8 (Ng et al. 1990) for an individual. For the average, we assume a geometric standard deviation of the mean of 1.5.

The controlling rate constant for the loss of ^{131}I from the pasture-cow-milk system is the effective rate of loss of ^{131}I from pasture λ. The nominal rate of loss independent of radioactive decay corresponds is 0.05 day[-1] (Thompson 1965). A somewhat arbitrary value of a geometric standard deviation of 1.5 is assigned to this value. The rate of loss due to decay is 0.0862 day[-1], and the effective rate of loss is 0.1362 day[-1]. The corresponding geometric mean is 0.137 day[-1] with a geometric standard deviation of 1.5.

The product of all values indicated above, with the exception of the number of persons, is 2.3 rad, which is taken to be the geometric mean of the dose to the population of the U.S. The associated geometric standard deviation of this number is 3.4. The arithmetic average that corresponds to these values (see, for example, Ng et al. 1990) is 5 rad, which when multiplied by the number of persons produces a value of the collective dose of 8×10^8 person-rad. The comparable number provided in the NCI report is 4×10^8 person-rad.

Further, the 5th and 95th percentile ranges of the collective dose according to the simplified scheme derived here are 5×10^7 and 3×10^9 person-rad.

References

Anspaugh, L. R.; McArthur, R. D. Inventory and distribution of residual ^{137}Cs on and near the Nevada Test Site (to be submitted).

Beck, H. A.; Helfer, I. K.; Bouville, A.; Dreicer, M. Estimates of fallout in the continental U.S. from Nevada weapons testing based on gummed-film monitoring data. Health Physics 59:565-576; 1990.

Church, B. W.; Wheeler, D. L.; Campbell, C. M.; Nutley, R. V.; Anspaugh, L. R. Overview of the Department of Energy's Off-Site Radiation Exposure Review Project (ORERP). Health Physics 59:503-510; 1990.

Funk & Wagnalls. The world almanac and book of facts 1995. In: Microsoft Bookshelf; 1994.

Garner, R. J.; Russell, R. S. Isotopes of iodine. In: Russell, R. S., ed.; Radioactivity and Human Diet. New York: Pergola Press; 1966.

Hicks, H. G. Calculation of the concentration of any radionuclide deposited on the ground by offsite fallout from a nuclear detonation. Health Physics 42:585-600; 1982.

Hicks, H. G. Additional calculations of radionuclide production following nuclear explosions and Pu isotopic ratios for Nevada Test Site Events. Health Physics 59:515-523; 1990.

Koranda, J. J. Agricultural factors affecting the daily intake of fresh fallout by dairy cows. Livermore, CA: Lawrence Livermore National Laboratory; Report UCRL-12479; 1965.

Ng, Y. C.; Anspaugh, L. R.; Cederwall, R. T. ORERP internal dose estimates for individuals. Health Physics 59:693-713; 1990.

Simon, S. L. An analysis of vegetation interception data pertaining to close-in weapons test fallout. Health Physics 59:619-626; 1990.

Thompson, S. E. Effective half-life of fallout radionuclides on plants with special emphasis on iodine-131. Livermore, CA: Lawrence Livermore National Laboratory; Report UCRL-12388; 1965.

Appendix D

Thyroid Cancer in Idaho
1970-1996

Christopher J. Johnson, MPH, Epidemiologist
Stacey L. Carson, ART, CTR, Director
Cancer Data Registry of Idaho
PO Box 1278
Boise, Idaho 83701-1278
(208) 338-5100 ext. 213 (phone)
(208) 338-7800 (FAX)

Following release of the National Cancer Institute's report of its study to assess Americans' exposure to radioactive iodine-131 from atmospheric nuclear bomb testing in the 1950s and 1960s at the Nevada Test Site, and pursuant to requests from the public, media, and health officials, staff at the Cancer Data Registry of Idaho (CDRI) conducted several analyses of thyroid cancer. This report describes the analyses of thyroid cancer incidence rates in Idaho, 1970-1996, and the ratio of female-to-male thyroid cancer cases by age group. Because four of the five counties in the United States with highest estimated exposure to iodine-131 from atmospheric nuclear bomb tests at the Nevada Test Site are located in Idaho (Blaine, Custer, Gem, and Lemhi), and public health services are delivered at the health district level, analyses were conducted at both the county and health district levels of geography (see Appendix for listing of counties by health district).

METHODS

Established in 1969, CDRI is a population-based cancer registry that collects incidence and survival data on cancer patients who reside in the state of Idaho at the time of diagnosis or who are diagnosed and/or treated for cancer in the state of Idaho. All cases of invasive thyroid cancer diagnosed among residents of the state of Idaho between January 1, 1970, and December 31, 1996, were included in these analyses.

Thyroid Cancer Incidence in Idaho, 1970-1996, by Birth Cohort and Overall

A combination of direct and indirect age adjustment was used to compare the incidence rates of thyroid cancer among geographic areas and by birth cohort. The following provides an overview of the steps taken in the analysis.

Step 1. The number of invasive cases of thyroid cancer diagnosed among residents of the state of Idaho, 1970-1996, were summed by sex, 5-year age group, county of residence, and year of diagnosis. The age groups were: 0-4, 5-9, 10-14, ..., 85+. Cases with missing information on sex, age at diagnosis, or county were not included in the analysis. If birth year was unknown, it was imputed from diagnosis date and estimated age at diagnosis. Birth cohort was defined by birth year: before 1948, 1948-1958, and after 1958. Given the years that atmospheric nuclear bomb tests were conducted at the Nevada Test Site, and the fact that children aged 0-5 are biologically most sensitive to iodine-131, persons born during the time period 1948-1958 are thought to have been at the risk of highest exposure. The other birth cohorts were chosen for comparison purposes.

Step 2. Population data were obtained from the U.S. Bureau of the Census. Data for 1970-1989 were available by sex and 5-year age-group. Data for 1990-1996 were available by sex and single-year age group. Single-year age group estimates were derived from 1970-1989 data by dividing the population in each 5-year age group into 5 equal parts. For example, if 1,000 persons were in the 5-year age group 15-19, 200 persons each were assigned to ages 15, 16, 17, 18, and 19. Separating the population data into single-year age groups was necessary to calculate age-adjusted rates by birth cohort.

Step 3. Three birth cohorts were defined, again based upon differences due to age in estimated exposure to iodine-131 from atmospheric nuclear bomb tests at the Nevada Test Site: before 1948, 1948-1958, and after 1958. Population data were estimated for each of the birth cohorts, by county, sex, and 5-year age group. Two age breaks were defined and incremented by year of study. The lower age break was defined as: **AGELOW = YEAR − 1958**, and year was varied from 1970-1996, the years for which CDRI has reliable statewide case information. The upper age break was defined as: **AGEHIGH = YEAR − 1948**, and year was again varied from 1970-1996. AGELOW and AGEHIGH give the (truncated) ages a person would have been in each year, 1970-1996, given that they were born in the period 1948-1958. Thus, a person born from 1948-1958 would have been assigned ages 12-22 in 1970, and between 32 and 42 in 1990, etc. Population data were assigned to birth cohorts depending on single-year age group and year (1970-1996). For example, for the year 1970, persons aged 10 were assigned to the after-1958 birth cohort; persons aged 20 were assigned to the 1948-1958 birth cohort; and persons aged 30 were assigned to the before-1948 birth cohort. Finally, population data were collapsed over age to yield estimates by birth cohort, county, sex, and 5-year age group.

Step 4. The age-adjusted rate of invasive thyroid cancer for the state of Idaho, 1970-1996, was calculated by the direct method, using the 1970 U.S. population as standard. Age-specific rates for the state of Idaho were calculated for use in direct age adjustment. The result was an age-adjusted incidence rate of thyroid cancer for the state of Idaho, 1970-1996, of 4.22 cases per 100,000 person-years.

Step 5. Age- and sex-specific rates for the state of Idaho for the time period 1970-1996 were calculated for use as standard rates in indirect age and sex adjustment.

Step 6. For all birth cohorts combined, the numbers of observed and expected cases were calculated for each health district and county. Expected cases were calculated by applying the age- and sex-specific rates for the state of Idaho to the population by age and sex in each health district and county. Two-tailed p-values comparing the number of observed and expected cases were calculated using the Poisson probability distribution. Observed and expected cases, and p-values, were calculated separately for males, females, and both sexes combined.

Step 7. The adjusted incidence rates for each health district and county were calculated as the standardized incidence ratio (observed/expected) multiplied by the age-adjusted rate from Step 4. For example, there were 314 cases observed, and 254.6 cases expected in Health District 4, yielding an adjusted incidence rate of (314/254.6) * 4.22 = 5.20 cases per 100,000 person-years.

Step 8. Age- and sex-specific rates for the state of Idaho, 1970-1996, were calculated by birth cohort for use in indirect age and sex adjustment by birth cohort. (Step 5 was repeated by birth cohort.)

Step 9. For each birth cohort, the number of observed and expected cases were calculated for each health district and county. Expected cases were calculated by applying the age- and sex-specific rates for the state of Idaho, by birth cohort (from Step 8), to the population by age and sex in each health district and county. Two-tailed p-values comparing the number of observed and expected cases were calculated using the Poisson probability distribution. Observed and expected cases, and p-values, were calculated separately for males, females, and both sexes combined, by birth cohort. Statistical significance was set at $\alpha = .05$.

Step 10. The adjusted rates for each health district and county were calculated by birth cohort as the standardized incidence ratio for that birth cohort (observed/expected) multiplied by the age-adjusted rate for all birth cohorts (from Step 4). For example, there were 87 cases observed and 69.4 expected in Health District 4 for the birth cohort 1948-1958, yielding an adjusted incidence rate of (87/69.4) * 4.22 = 5.29. The overall rate from Step 4 was used as the reference rate in order to facilitate comparisons across birth cohort for individual health districts and counties, and to allow comparisons among health districts and counties within birth cohort. See Box D.1 for a description of the Idaho population by health districts.

Step 11. The adjusted rates for the state of Idaho were calculated by birth cohort using the standardized incidence ratio for that birth cohort (observed/expected), with expected cases based upon age- and sex-specific rates for all birth cohorts

---

BOX D.1
Description of Idaho Population

The population of the state of Idaho in 1996 was estimated to be 1,189,251, made up of 594,604 males and 594,647 females (U.S. Bureau of the Census). Idaho comprises 44 counties grouped into seven health districts. The composition of the health districts and population estimates by sex are shown below:

District	Counties	Males	Females
District 1	Benewah, Bonner, Boundary, Kootenai, Shoshone	80,792	81,548
District 2	Clearwater, Idaho, Latah, Lewis, Nez Perce	49,574	48,568
District 3	Adams, Canyon, Gem, Owyhee, Payette, Washington	84,753	85,602
District 4	Ada, Boise, Elmore, Valley	147,684	149,119
District 5	Blaine, Camas, Cassia, Gooding, Jerome, Lincoln, Minidoka, Twin Falls	77,788	77,139
District 6	Bannock, Bear Lake, Bingham, Butte, Caribou, Franklin, Oneida, Power	77,425	77,227
District 7	Bonneville, Clark, Custer, Fremont, Jefferson, Lemhi, Madison, Teton	76,588	75,444

(from Step 5). For example, there were 271 cases observed and 256.9 expected in the state of Idaho, 1970-1996, among the 1948-1958 cohort, yielding an adjusted incidence rate of (271/256.9) * 4.22 = 4.45 cases per 100,000 person-years. The state of Idaho rates by birth cohort were designed to be compared to each other, and to the overall state of Idaho rate of 4.22 cases per 100,000 person-years.

Ratio of Female-to-Male Thyroid Cancer Cases by Age Group and Birth Cohort

The overall ratio of thyroid cancer cases among females versus males differed by birth cohort, with a ratio of 2.7 in the before-1948 cohort, 5.2 in the 1948-1958 cohort, and 5.3 in the after-1958 cohort. In order to examine if the differences in female-to-male ratios by birth cohort were an artifact of differing age-specific rates by sex, cumulative ratios of female-to-male cases were calculated by age group. For all invasive cases of thyroid cancer diagnosed among Idaho residents, 1970-1996, the numbers of cases were summed separately for males and females by 5-year age group and birth cohort. The cumulative ratios of female-to-male cases were calculated by 5-year age group and birth cohort. For example, for the age group 35-39, the cumulative female-to-male ratio in the before-1948 birth cohort was 5.0 (75 cases among females aged 39 years and younger, and 15 cases among males aged 39 years and younger).

RESULTS

The overall age-adjusted incidence rate of invasive thyroid cancer in Idaho, 1970-1996, was 4.22 cases per 100,000 person-years (see Table D.1). Incidence rates varied by geographic location, ranging from 3.28 cases per 100,000 person-years in Health District 2 to 5.20 cases per 100,000 person-years in Health District 4. There were significantly more cases of invasive thyroid cancer diagnosed among residents of Health District 4 than expected based upon rates in the state of Idaho (314 observed, 254.6 expected, p<.001), and the number of observed cases was higher than expected for both males and females. Ada County had significantly more cases of invasive thyroid cancer than expected based upon rates in the state of Idaho (276 observed, 221.4 expected, p<.001). None of the four Idaho counties with highest estimated exposure to iodine-131 showed an elevation in thyroid cancer cases from 1970-1996.

Among the birth cohort born before 1948, the incidence rate of invasive thyroid cancer, 1970-1996, was 4.10 cases per 100,000 person-years (see Table D.2). There were significantly more cases observed than expected in Health District 4 and Ada County. None of the four Idaho counties with highest estimated exposure to iodine-131 showed an elevation in thyroid cancer cases from 1970-1996 in the birth cohort born before 1948.

Among the birth cohort born 1948-1958, the incidence rate of invasive thyroid cancer, 1970-1996, was 4.45 cases per 100,000 person-years (see Table D.3). There were significantly more cases observed than expected in Health District 4 and Ada County. None of the four Idaho counties with highest estimated exposure to iodine-131 showed an elevation in thyroid cancer cases from 1970-1996 in the birth cohort born 1948-1958. Although the incidence rate of invasive thyroid cancer, 1970-1996, was highest for the birth cohort born 1948-1958, the number of cases observed was not statistically significantly different from that expected based upon rates for all birth cohorts.

Among the birth cohort born after 1958, the incidence rate of invasive thyroid cancer, 1970-1996, was 4.35 cases per 100,000 person-years (see Table D.4). There were significantly more cases observed than expected in Elmore County. None of the four counties with highest estimated exposure to iodine-131 showed an elevation in thyroid cancer cases from 1970-1996 in the birth cohort born after 1958.

Regarding the female-to-male ratios for invasive thyroid cancer cases, the differences in the overall female-to-male ratios by birth cohort (see Table D.5) appear to be due to the higher age-specific thyroid cancer incidence rates in younger females as compared with younger males. In all three birth cohorts, the cumulative age-specific ratios were similar for the age groups 25-29, 30-34, and 35-39 (the only age groups for which comparisons are available across all three birth cohorts, as CDRI has reliable statewide cancer incidence data since 1970).

SUMMARY AND CONCLUSIONS

Thyroid cancer is relatively rare among all cancers, accounting for less than 2% of invasive cases in Idaho in 1996. The age-adjusted incidence rate of invasive thyroid cancer in Idaho, 1970-1996, was 4.22 cases per 100,000 person-years. In comparison, the Surveillance, Epidemiology, and End Results (SEER) rate for whites, 1973-1995, was 4.39 cases per 100,000 person-years. CDRI investigated thyroid cancer incidence in three birth cohorts to explore the relationship between age at the time of iodine-131 release from atmospheric nuclear bomb tests at the Nevada Test Site and thyroid cancer incidence. The number of invasive thyroid cancer cases in the state of Idaho, 1970-1996, was not statistically significantly higher than expected, based upon overall rates, for any of the three birth cohorts. Within each birth cohort, and for all cohorts combined, variation existed among health districts and counties in the incidence of thyroid cancer, with more marked variation observed among geographic areas with smaller populations.

There are several limitations of the data that may have influenced the results of the analyses. The accuracy of the estimated incidence rates assumes similar case

TABLE D.1 Invasive Thyroid Cancer in Idaho, 1970-1996, Among All Birth Cohorts

Residence	Incidence Rate	All Cases			Male Cases			Female Cases		
		Observed	Expected	P-Value	Observed	Expected	P-Value	Observed	Expected	P-Value
STATE OF IDAHO	4.22	1,119	n/a	n/a	256	n/a	n/a	863	n/a	n/a
HEALTH DISTRICT 1	3.60	127	148.9	0.074	26	35.2	0.132	101	113.7	0.250
HEALTH DISTRICT 2	3.28	85	109.4	0.018	18	25.8	0.141	67	83.6	0.072
HEALTH DISTRICT 3	4.34	165	160.6	0.751	39	37.4	0.838	126	123.2	0.826
HEALTH DISTRICT 4	5.20	314	254.6	0.000	70	55.5	0.067	244	199.2	0.002
HEALTH DISTRICT 5	4.59	171	157.2	0.291	37	36.9	1.000	134	120.3	0.232
HEALTH DISTRICT 6	3.52	126	151.0	0.042	34	34.3	1.000	92	116.7	0.021
HEALTH DISTRICT 7	4.03	131	137.2	0.631	32	30.9	0.897	99	106.3	0.515
ADA	5.26	276	221.4	0.000	63	47.7	0.039	213	173.7	0.004
ADAMS	4.13	4	4.1	1.000	2	1.1	0.570	2	3.0	0.834
BANNOCK	4.19	72	72.5	1.000	17	15.8	0.828	55	56.7	0.894
BEAR LAKE	4.64	8	7.3	0.887	2	1.8	1.000	6	5.5	0.949
BENEWAH	2.67	6	9.5	0.333	1	2.4	0.624	5	7.1	0.578
BINGHAM	3.29	30	38.5	0.190	11	8.9	0.553	19	29.7	0.050
BLAINE	4.80	16	14.1	0.674	4	3.2	0.783	12	10.9	0.815
BOISE	4.42	4	3.8	1.000	2	1.0	0.510	2	2.8	0.916
BONNER	3.40	25	31.0	0.322	6	7.6	0.733	19	23.4	0.424
BONNEVILLE	4.31	75	73.5	0.888	14	16.3	0.685	61	57.2	0.648
BOUNDARY	5.13	11	9.0	0.598	4	2.2	0.362	7	6.8	1.000
BUTTE	2.38	2	3.5	0.626	1	0.9	1.000	1	2.6	0.519
CAMAS	0.00	—	0.9	0.793	—	0.2	1.000	—	0.7	1.000
CANYON	4.67	113	102.1	0.305	28	23.1	0.354	85	79.0	0.532
CARIBOU	2.08	4	8.1	0.184	—	1.9	0.299	4	6.2	0.510
CASSIA	3.37	17	21.3	0.421	5	5.0	1.000	12	16.3	0.348
CLARK	0.00	—	0.9	0.821	—	0.3	1.000	—	0.6	1.000
CLEARWATER	3.93	11	11.8	0.969	2	3.0	0.843	9	8.8	1.000
CUSTER	7.91	9	4.8	0.112	5	1.2	0.017	4	3.6	0.957

County										
ELMORE	4.40	23	22.0	0.894	2	5.0	0.251	21	17.1	0.396
FRANKLIN	0.88	2	9.6	0.008	1	2.3	0.641	1	7.3	0.012
FREMONT	3.34	9	11.4	0.600	—	2.7	0.130	9	8.7	0.996
GEM	4.66	16	14.5	0.757	3	3.5	1.000	13	10.9	0.607
GOODING	3.87	13	14.2	0.888	3	3.6	1.000	10	10.6	1.000
IDAHO	2.47	10	17.1	0.094	1	4.3	0.140	9	12.8	0.363
JEFFERSON	2.90	11	16.0	0.255	4	3.8	1.000	7	12.2	0.164
JEROME	4.85	20	17.4	0.594	4	4.1	1.000	16	13.3	0.519
KOOTENAI	3.53	66	78.9	0.156	11	18.1	0.105	55	60.8	0.501
LATAH	3.40	27	33.5	0.295	6	7.4	0.771	21	26.1	0.372
LEMHI	4.29	9	8.9	1.000	2	2.2	1.000	7	6.6	0.989
LEWIS	0.87	1	4.8	0.092	—	1.2	0.580	1	3.6	0.249
LINCOLN	3.13	3	4.0	0.850	—	1.0	0.713	3	3.0	1.000
MADISON	4.17	18	18.2	1.000	7	3.5	0.134	11	14.7	0.409
MINIDOKA	4.51	23	21.5	0.806	1	5.1	0.074	22	16.4	0.216
NEZ PERCE	3.61	36	42.1	0.394	9	9.7	0.980	27	32.3	0.400
ONEIDA	3.21	3	3.9	0.889	—	1.0	0.749	3	3.0	1.000
OWYHEE	2.76	6	9.2	0.381	1	2.3	0.647	5	6.9	0.639
PAYETTE	3.21	15	19.7	0.344	1	4.7	0.106	14	15.0	0.924
POWER	2.82	5	7.5	0.486	2	1.7	1.000	3	5.7	0.352
SHOSHONE	3.92	19	20.4	0.863	4	4.9	0.902	15	15.5	1.000
TETON	0.00	—	3.6	0.053	—	0.9	0.808	—	2.7	0.131
TWIN FALLS	5.22	79	63.8	0.073	20	14.7	0.216	59	49.1	0.187
VALLEY	6.33	11	7.3	0.248	3	1.8	0.531	8	5.5	0.393
WASHINGTON	4.20	11	11.0	1.000	4	2.7	0.581	7	8.3	0.817

NOTES: The incidence rate for the state of Idaho is age adjusted to the standard 1970 U.S. population using the direct method. The incidence rates for the other geographic areas are the products of the standardized incidence ratios and the state age-adjusted rate. Expected cases are based upon age- and sex-specific rates for the state of Idaho. P-values compare observed and expected cases, are two-tailed, based upon the Poisson probability distribution.

TABLE D.2 Invasive Thyroid Cancer in Idaho, 1970-1996, Among Birth Cohort Born Before 1948

Residence	Incidence Rate	All Cases			Male Cases			Female Cases		
		Observed	Expected	P-Value	Observed	Expected	P-Value	Observed	Expected	P-Value
STATE OF IDAHO	4.10	679	698.3	0.480	185	186.5	0.953	494	511.8	0.447
HEALTH DISTRICT 1	3.55	79	93.8	0.133	21	26.4	0.346	58	67.5	0.273
HEALTH DISTRICT 2	3.08	50	68.5	0.024	13	19.1	0.190	37	49.4	0.082
HEALTH DISTRICT 3	4.32	106	103.6	0.841	31	28.4	0.669	75	75.3	1.000
HEALTH DISTRICT 4	5.13	173	142.3	0.014	44	37.3	0.306	129	105.0	0.026
HEALTH DISTRICT 5	4.54	108	100.3	0.468	25	27.6	0.716	83	72.8	0.255
HEALTH DISTRICT 6	3.77	81	90.7	0.334	27	24.6	0.682	54	66.1	0.147
HEALTH DISTRICT 7	4.34	82	79.8	0.837	24	21.8	0.688	58	58.0	1.000
ADA	5.35	157	123.9	0.005	40	32.2	0.201	117	91.8	0.013
ADAMS	6.22	4	2.7	0.578	2	0.8	0.403	2	1.9	1.000
BANNOCK	4.41	43	41.2	0.815	13	10.9	0.595	30	30.3	1.000
BEAR LAKE	5.10	6	5.0	0.755	2	1.4	0.805	4	3.6	0.960
BENEWAH	3.45	5	6.1	0.854	1	1.8	0.915	4	4.3	1.000
BINGHAM	4.13	23	23.5	1.000	10	6.4	0.230	13	17.1	0.393
BLAINE	3.60	6	7.0	0.888	—	1.9	0.288	6	5.1	0.803
BOISE	5.52	3	2.3	0.805	2	0.7	0.319	1	1.6	1.000
BONNER	3.20	15	19.8	0.339	5	5.7	0.983	10	14.0	0.347
BONNEVILLE	4.35	44	42.7	0.882	8	11.3	0.420	36	31.4	0.459
BOUNDARY	6.62	9	5.7	0.254	3	1.6	0.458	6	4.1	0.459
BUTTE	0.00	—	2.4	0.187	—	0.7	0.990	—	1.7	0.377
CAMAS	0.00	—	0.6	1.000	—	0.2	1.000	—	0.4	1.000
CANYON	4.63	70	63.8	0.472	21	17.1	0.397	49	46.8	0.783
CARIBOU	1.65	2	5.1	0.232	—	1.4	0.494	2	3.7	0.567
CASSIA	3.13	10	13.5	0.422	4	3.7	1.000	6	9.8	0.286
CLARK	0.00	—	0.6	1.000	—	0.2	1.000	—	0.4	1.000
CLEARWATER	3.72	7	7.9	0.922	2	2.3	1.000	5	5.6	1.000
CUSTER	6.79	5	3.1	0.406	3	0.9	0.140	2	2.2	1.000

County	Incidence rate	Observed	Expected	P-value	Observed	Expected	P-value	Observed	Expected	P-value
ELMORE	2.56	7	11.6	0.222	—	3.1	0.091	7	8.5	0.781
FRANKLIN	0.00	—	6.4	0.003	—	1.8	0.330	—	4.6	0.021
FREMONT	2.92	5	7.2	0.543	—	2.0	0.261	5	5.2	1.000
GEM	3.82	9	9.9	0.930	3	2.8	1.000	6	7.1	0.864
GOODING	4.28	10	9.9	1.000	3	2.8	1.000	7	7.0	1.000
IDAHO	2.55	7	11.6	0.216	1	3.4	0.298	6	8.2	0.573
JEFFERSON	3.48	8	9.7	0.735	4	2.8	0.598	4	6.9	0.357
JEROME	5.63	15	11.2	0.329	4	3.1	0.770	11	8.1	0.389
KOOTENAI	3.32	38	48.4	0.148	8	13.3	0.176	30	35.1	0.446
LATAH	2.83	12	17.9	0.192	3	4.9	0.562	9	13.0	0.333
LEMHI	4.20	6	6.0	1.000	2	1.8	1.000	4	4.3	1.000
LEWIS	1.24	1	3.4	0.292	—	1.0	0.741	1	2.4	0.610
LINCOLN	1.55	1	2.7	0.490	—	0.8	0.898	1	1.9	0.856
MADISON	7.13	14	8.3	0.087	7	2.2	0.014	7	6.1	0.821
MINIDOKA	3.71	12	13.6	0.790	—	3.8	0.044	12	9.8	0.568
NEZ PERCE	3.51	23	27.6	0.440	7	7.5	1.000	16	20.1	0.425
ONEIDA	3.02	2	2.8	0.943	—	0.8	0.908	2	2.0	1.000
OWYHEE	3.55	5	6.0	0.907	1	1.8	0.948	4	4.2	1.000
PAYETTE	3.18	10	13.3	0.457	1	3.7	0.237	9	9.6	1.000
POWER	4.73	5	4.5	0.923	2	1.2	0.709	3	3.2	1.000
SHOSHONE	3.66	12	13.9	0.746	4	3.9	1.000	8	10.0	0.673
TETON	0.00	—	2.2	0.225	—	0.6	1.000	—	1.5	0.429
TWIN FALLS	5.47	54	41.7	0.076	14	11.2	0.467	40	30.5	0.114
VALLEY	5.66	6	4.5	0.584	2	1.3	0.743	4	3.2	0.784
WASHINGTON	4.27	8	7.9	1.000	3	2.2	0.766	5	5.7	0.997

NOTES: The incidence rate for the state of Idaho is the product of the age adjusted rate for all birth cohorts using the 1970 standard U.S. population (direct age adjustment) and the standardized incidence ratio for this birth cohort compared with all birth cohorts (indirect age adjustment). The incidence rates for the other geographic areas are the products of the standardized incidence ratios for this cohort and the state age-adjusted rate. Expected cases for geography other than state are based upon age- and sex-specific rates for the state of Idaho for this cohort. P-values compare observed and expected cases, are two-tailed, based upon the Poisson probability distribution.

TABLE D.3 Invasive Thyroid Cancer in Idaho, 1970-1996, Among Birth Cohort Born 1948-1958

Residence	Incidence Rate	All Cases			Male Cases			Female Cases		
		Observed	Expected	P-Value	Observed	Expected	P-Value	Observed	Expected	P-Value
STATE OF IDAHO	4.45	271	256.9	0.394	44	45.2	0.932	227	211.6	0.307
HEALTH DISTRICT 1	3.18	27	35.8	0.156	4	5.7	0.654	23	30.1	0.223
HEALTH DISTRICT 2	4.10	24	24.7	0.992	3	4.2	0.804	21	20.6	0.980
HEALTH DISTRICT 3	4.35	36	35.0	0.905	6	5.5	0.950	30	29.4	0.966
HEALTH DISTRICT 4	5.29	87	69.4	0.046	16	11.4	0.227	71	58.0	0.109
HEALTH DISTRICT 5	4.38	37	35.6	0.865	5	5.8	0.948	32	29.8	0.737
HEALTH DISTRICT 6	3.44	30	36.8	0.297	7	6.0	0.775	23	30.8	0.177
HEALTH DISTRICT 7	3.76	30	33.7	0.599	3	5.5	0.413	27	28.2	0.919
ADA	5.39	77	60.3	0.043	13	9.7	0.361	64	50.6	0.078
ADAMS	0.00	—	0.9	0.824	—	0.1	1.000	—	0.7	0.953
BANNOCK	4.20	19	19.1	1.000	4	3.1	0.741	15	16.0	0.930
BEAR LAKE	2.94	1	1.4	1.000	—	0.2	1.000	1	1.2	1.000
BENEWAH	0.00	—	2.2	0.228	—	0.4	1.000	—	1.8	0.329
BINGHAM	1.39	3	9.1	0.040	1	1.5	1.000	2	7.6	0.036
BLAINE	3.59	4	4.7	0.987	—	0.8	0.887	4	3.9	1.000
BOISE	4.15	1	1.0	1.000	—	0.2	1.000	1	0.8	1.000
BONNER	3.90	7	7.6	1.000	1	1.2	1.000	6	6.3	1.000
BONNEVILLE	4.28	19	18.7	1.000	3	3.0	1.000	16	15.7	1.000
BOUNDARY	3.96	2	2.1	1.000	1	0.4	0.594	1	1.8	0.940
BUTTE	11.27	2	0.7	0.346	1	0.1	0.240	1	0.6	0.925
CAMAS	0.00	—	0.2	1.000	—	0.0	1.000	—	0.2	1.000
CANYON	4.53	25	23.3	0.774	5	3.7	0.613	20	19.6	0.988
CARIBOU	4.43	2	1.9	1.000	—	0.3	1.000	2	1.6	0.946
CASSIA	4.46	5	4.7	1.000	1	0.8	1.000	4	4.0	1.000
CLARK	0.00	—	0.2	1.000	—	0.0	1.000	—	0.1	1.000
CLEARWATER	6.78	4	2.5	0.480	—	0.4	1.000	4	2.1	0.311
CUSTER	7.50	2	1.1	0.621	—	0.2	1.000	2	0.9	0.480

ELMORE	3.46	5	6.1	0.862	2	1.2	0.645	3	4.9	0.549
FRANKLIN	4.42	2	1.9	1.000	1	0.3	0.550	1	1.6	1.000
FREMONT	3.37	2	2.5	1.000	—	0.4	1.000	2	2.1	1.000
GEM	7.37	5	2.9	0.325	—	0.4	1.000	5	2.4	0.196
GOODING	4.64	3	2.7	1.000	—	0.5	1.000	3	2.3	0.793
IDAHO	2.38	2	3.5	0.626	—	0.6	1.000	2	2.9	0.879
JEFFERSON	2.23	2	3.8	0.541	—	0.6	1.000	2	3.2	0.777
JEROME	4.39	4	3.8	1.000	—	0.6	1.000	4	3.2	0.805
KOOTENAI	3.01	14	19.6	0.240	2	3.1	0.818	12	16.6	0.316
LATAH	5.28	11	8.8	0.541	2	1.6	0.918	9	7.2	0.608
LEMHI	2.25	1	1.9	0.880	—	0.3	1.000	1	1.6	1.000
LEWIS	0.00	—	0.9	0.803	—	0.2	1.000	—	0.8	0.939
LINCOLN	0.00	—	0.8	0.858	—	0.1	1.000	—	0.7	0.997
MADISON	3.70	4	4.6	1.000	—	0.7	0.983	4	3.9	1.000
MINIDOKA	7.05	8	4.8	0.224	—	0.8	0.900	8	4.0	0.101
NEZ PERCE	3.29	7	9.0	0.654	1	1.4	1.000	6	7.6	0.742
ONEIDA	5.84	1	0.7	1.000	—	0.1	1.000	1	0.6	0.909
OWYHEE	2.14	1	2.0	0.829	—	0.3	1.000	1	1.6	1.000
PAYETTE	2.11	2	4.0	0.477	—	0.6	1.000	2	3.4	0.686
POWER	0.00	—	1.9	0.301	—	0.3	1.000	—	1.6	0.411
SHOSHONE	3.94	4	4.3	1.000	—	0.7	1.000	4	3.6	0.970
TETON	0.00	—	0.9	0.821	—	0.2	1.000	—	0.7	0.960
TWIN FALLS	3.98	13	13.8	0.971	4	2.2	0.358	9	11.6	0.556
VALLEY	8.51	4	2.0	0.280	1	0.3	0.569	3	1.6	0.458
WASHINGTON	6.40	3	2.0	0.635	1	0.3	0.544	2	1.7	0.990

NOTES: The incidence rate for the state of Idaho is the product of the age adjusted rate for all birth cohorts using the 1970 standard U.S. population (direct age adjustment) and the standardized incidence ratio for this birth cohort compared with all birth cohorts (indirect age adjustment). The incidence rates for the other geographic areas are the products of the standardized incidence ratios for this cohort and the state age-adjusted rate. Expected cases for geography other than state are based upon age- and sex-specific rates for the state of Idaho for this cohort. P-values compare observed and expected cases, are two-tailed, based upon the Poisson probability distribution.

TABLE D.4 Invasive Thyroid Cancer in Idaho, 1970-1996, Among Birth Cohort Born After 1958

Residence	Incidence Rate	All Cases			Male Cases			Female Cases		
		Observed	Expected	P-Value	Observed	Expected	P-Value	Observed	Expected	P-Value
STATE OF IDAHO	4.35	169	163.9	0.708	27	24.3	0.634	142	139.6	0.859
HEALTH DISTRICT 1	4.57	21	19.4	0.774	1	3.1	0.367	20	16.3	0.416
HEALTH DISTRICT 2	2.92	11	15.9	0.265	2	2.5	1.000	9	13.4	0.282
HEALTH DISTRICT 3	4.41	23	22.0	0.887	2	3.5	0.631	21	18.5	0.615
HEALTH DISTRICT 4	5.20	54	43.8	0.152	10	6.9	0.329	44	36.9	0.279
HEALTH DISTRICT 5	5.20	26	21.1	0.336	7	3.5	0.137	19	17.6	0.793
HEALTH DISTRICT 6	2.74	15	23.1	0.099	—	3.7	0.050	15	19.4	0.375
HEALTH DISTRICT 7	3.39	19	23.6	0.398	5	3.7	0.621	14	20.0	0.213
ADA	4.66	42	38.1	0.565	10	5.9	0.159	32	32.1	1.000
ADAMS	0.00	—	0.5	1.000	—	0.1	1.000	—	0.4	1.000
BANNOCK	3.49	10	12.1	0.678	—	1.8	0.321	10	10.2	1.000
BEAR LAKE	4.92	1	0.9	1.000	—	0.1	1.000	1	0.7	1.000
BENEWAH	3.60	1	1.2	1.000	—	0.2	1.000	1	1.0	1.000
BINGHAM	2.89	4	5.9	0.611	4	1.0	0.747	4	4.9	0.928
BLAINE	10.53	6	2.4	0.072	—	0.4	0.002	2	2.0	1.000
BOISE	0.00	—	0.5	1.000	—	0.1	1.000	—	0.4	1.000
BONNER	3.41	3	3.7	0.984	3	0.6	1.000	3	3.1	1.000
BONNEVILLE	4.23	12	12.0	1.000	—	2.0	0.648	9	10.0	0.922
BOUNDARY	0.00	—	1.2	0.621	—	0.2	1.000	—	1.0	0.759
BUTTE	0.00	—	0.4	1.000	—	0.1	1.000	—	0.3	1.000
CAMAS	0.00	—	0.1	1.000	—	0.0	1.000	—	0.1	1.000
CANYON	5.04	18	15.1	0.514	2	2.4	1.000	16	12.7	0.420
CARIBOU	0.00	—	1.1	0.693	—	0.2	1.000	—	0.9	0.828
CASSIA	2.86	2	3.0	0.867	—	0.5	1.000	2	2.5	1.000
CLARK	0.00	—	0.1	1.000	—	0.0	1.000	—	0.1	1.000
CLEARWATER	0.00	—	1.3	0.566	—	0.2	1.000	—	1.0	0.713
CUSTER	14.82	2	0.6	0.224	2	0.1	0.009	—	0.5	1.000

ELMORE	10.58	11	4.4	0.011	—	0.8	0.918	11	3.6	0.003
FRANKLIN	0.00	—	1.3	0.538	—	0.2	1.000	—	1.1	0.671
FREMONT	5.22	2	1.6	0.961	—	0.3	1.000	2	1.3	0.767
GEM	5.11	2	1.7	0.984	—	0.3	1.000	2	1.4	0.796
GOODING	0.00	—	1.6	0.405	—	0.3	1.000	—	1.3	0.539
IDAHO	2.23	1	1.9	0.870	—	0.4	1.000	1	1.6	1.000
JEFFERSON	1.69	1	2.5	0.574	—	0.4	1.000	1	2.1	0.771
JEROME	1.83	1	2.3	0.659	—	0.4	1.000	1	1.9	0.858
KOOTENAI	5.26	14	11.2	0.480	1	1.8	0.950	13	9.5	0.321
LATAH	2.47	4	6.8	0.376	1	1.0	1.000	3	5.8	0.332
LEMHI	9.16	2	0.9	0.471	—	0.1	1.000	2	0.8	0.365
LEWIS	0.00	—	0.5	1.000	—	0.1	1.000	—	0.4	1.000
LINCOLN	18.17	2	0.5	0.159	—	0.1	1.000	2	0.4	0.115
MADISON	0.00	—	5.4	0.009	—	0.6	1.000	—	4.8	0.017
MINIDOKA	4.16	3	3.0	1.000	1	0.5	0.797	2	2.5	1.000
NEZ PERCE	4.69	6	5.4	0.906	1	0.8	1.000	5	4.6	0.957
ONEIDA	0.00	—	0.4	1.000	—	0.1	1.000	—	0.4	1.000
OWYHEE	0.00	—	1.3	0.572	—	0.2	1.000	—	1.0	0.723
PAYETTE	5.22	3	2.4	0.873	—	0.4	1.000	3	2.0	0.668
POWER	0.00	—	1.1	0.643	—	0.2	1.000	—	0.9	0.775
SHOSHONE	5.98	3	2.1	0.710	—	0.3	1.000	3	1.8	0.525
TETON	0.00	—	0.6	1.000	—	0.1	1.000	—	0.5	1.000
TWIN FALLS	6.16	12	8.2	0.258	2	1.3	0.771	10	6.9	0.318
VALLEY	4.82	1	0.9	1.000	—	0.1	1.000	1	0.7	1.000
WASHINGTON	0.00	—	1.1	0.645	—	0.2	1.000	—	1.0	0.772

NOTES: The incidence rate for the state of Idaho is the product of the age adjusted rate for all birth cohorts using the 1970 standard U.S. population (direct age adjustment) and the standardized incidence ratio for this birth cohort compared with all birth cohorts (indirect age adjustment). The incidence rates for the other geographic areas are the products of the standardized incidence ratios for this cohort and the state age-adjusted rate. Expected cases for geography other than state are based upon age- and sex-specific rates for the state of Idaho for this cohort. P-values compare observed and expected cases, are two-tailed, based upon the Poisson probability distribution.

TABLE D.5 Ratio of Female-to-Male Invasive Thyroid Cancer Cases by Age Group and Birth Cohort

Age Group	Birth Cohort Born Before 1948			Birth Cohort Born 1948-1958			Birth Cohort Born After 1958		
	Female Cases	Male Cases	Cumulative F:M Ratio	Female Cases	Male Cases	Cumulative F:M Ratio	Female Cases	Male Cases	Cumulative F:M Ratio
0-4							0	1	0.0
5-9							2	0	2.0
10-14							5	3	1.8
15-19				7	2	3.5	21	3	4.0
20-24				22	5	4.1	42	3	7.0
25-29	8	1	8.0	37	3	6.6	44	9	6.0
30-34	33	4	8.2	58	9	6.5	25	5	5.8
35-39	34	10	5.0	50	14	5.3	3	3	5.3
40-44	52	20	3.6	42	11	4.9			
45-49	70	15	3.9	11	0	5.2			
50-54	55	23	3.5						
55-59	60	20	3.4						
60-64	50	29	3.0						
65-69	37	17	2.9						
70-74	37	22	2.7						
75-79	30	12	2.7						
80-84	16	10	2.6						
85+	12	2	2.7						
Total	494	185	2.7	227	44	5.2	142	27	5.3

ascertainment rates across geographic areas, and depends upon the accuracy of the population estimates generated by the U.S. Bureau of the Census and apportionment of 5-year age group data into five equal single-year groups. The analyses did not account for in- or out-migration of cases. Cases among previous Idaho residents who moved to another state prior to diagnosis were not accounted for, but cases among persons who previously resided in other states, but moved to Idaho prior to diagnosis, were counted as Idaho cases. This same limitation exists at the county level for county-specific analyses. Many of the statistics, particularly for the county analyses by birth cohort, are based upon small numbers of cases. Incidence rates based upon 10 or fewer cases (numerator) should be interpreted with caution.

None of the four Idaho counties with highest estimated exposure to iodine-131 showed an elevation in thyroid cancer cases from 1970-1996 in the birth cohort born during 1948-1958, which was estimated to have received the highest exposure to iodine-131. Although the overall female-to-male thyroid case ratios differed by birth cohort, this result was confounded by age, as the cumulative age-specific ratios are similar in each birth cohort.

Appendix E

Applicable Radiation Exposure Standards and Guides: Past and Present

I. Introduction

According to the "Statement of Task," the committee is "to provide information that will enable the DHHS to educate and inform members of the public, especially those likely to have been most heavily exposed at the most vulnerable ages, as to what the NCI estimates and their attendant uncertainties mean for the individual, what the risks are relative to other environmental risks, and what types of appropriate actions they should take."

In responding to this request, it would appear beneficial to review the radiation protection standards, guides, and regulations that applied during the period from 1950 to 1970 (as well as in the subsequent years), and to review and evaluate the thyroid doses that occurred during the I-131 fallout in comparison to these limits and recommendations.

II. Relevant Standards and Guides—National Council on Radiation Protection and Measurements

1. Handbook 52

In 1953, the National Council on Radiation Protection (NCRP) published Handbook 52, "Maximum Permissible Amounts of Radioisotopes in the Human Body and Maximum Permissible Concentrations in Air and Water" (NCRP 1953). Although the accompanying standards were developed primarily for application to radiation workers, because few members of the public were anticipated to be exposed to artificial sources of radiation exposure (beyond the use of radiation in the healing arts) the NCRP also stated that the permissible concentrations were "for use beyond the control area." (Table 2, page 11).

The recommended limit for I-131 in drinking water was 3×10^{-5} µCi per milliliter (Table 3, page 15). This corresponds to a daily intake limit of 6.6×10^{-2} µCi which, if consumed over an entire year, would result in an estimated dose rate to the thyroid of 0.3 rem per week, or 15 rem per year (Section E, page 12).

2. Handbook 59

In a report issued in 1954, the NCRP (1954) recommended that

" ... in the course of their normal activities, protective measures be taken to make sure that no minor actually receives radiation at a weekly rate higher than one-tenth the respective permissible weekly doses for the critical organs" (page 78).

3. Addendum to Handbook 59

In 1957, the NCRP issued an addendum to Handbook 59. This addendum repeated the NCRP recommendation for an occupational dose rate limit of 15 rem per year, but the maximum accumulated dose was limited to 5 times the number of years beyond age 18 (NCRP, 1957, paragraph 1, page 2). This addendum also included new recommendations for limiting the genetic dose to the population (paragraph 5, page 3), and stated, in terms of internal emitters, that:

"For individuals outside of the controlled area, the maximum permissible concentrations should be one-tenth of those for occupational exposures" (paragraph 6, page 3).

The NCRP also recommended:

An average per capita dose limit of 0.5 rem per year for "persons outside of controlled areas, but exposed to radiation from a controlled area" (paragraph 12, page 4).

It should be noted, however, that this dose rate limit was based on the assumption that:

" ... the total integrated RBE dose received by radiation workers will be small in comparison with the integrated RBE dose of the whole population" (paragraph 12, page 4).

In addition, for purposes of controlling potential genetic effects, it was further assumed that:

" ... persons outside of controlled areas, but exposed to radiation from a controlled area, constitute only a small portion of the whole population."

4. Handbook 69

In 1959, the NCRP issued Handbook 69 (NCRP 1959) which incorporated new data and methods for estimating the dose from internally deposited radionuclides. In the Handbook the NCRP stated its recommendations as follows:

"The radiation or radioactive materials outside a controlled area, attributable to normal operations within the controlled area, shall be such that it is improbable that any individual will receive a dose of more than 0.5 rem in any 1 year from external radiation" (Section 2.4, page 6).

and:

"The maximum permissible average body burden of radionuclides in persons outside the controlled area and attributable to the operations within the controlled area shall not exceed one-tenth that for radiation workers" (Section 2.4, page 6).

At the same time, the NCRP repeated the limit on cumulative dose to radiation workers, namely, that the cumulative dose "shall not exceed 5 rems multiplied by the number of years beyond age 18."

At the same time, however, the NCRP increased the permissible dose rate to the thyroid of radiation workers to 30 rem per year (Section 2.1, pages 4-5).

With the application of the one-tenth factor, the limiting concentration of I-131 in drinking water was reduced to 2×10^{-6} µCi per milliliter, as contrasted to the 3×10^{-5} µCi per milliliter, as given in Handbook 52.

In support of these recommendations, the NCRP stated:

"The maximum permissible dose and the maximum permissible concentrations of radionuclides as recommended ... are primarily for the purpose of keeping the average dose to the whole population as low as reasonably achievable, and not because of the likelihood of specific injury to the individual."

5. Report of Ad Hoc Committee

In 1960, the NCRP published a report of an Ad Hoc Committee established "to re-examine the problem of exposure of the population to man-made radiations from the point of view of somatic effects as distinct from genetic effects" (page 482). This review was undertaken in specific response to "the widespread public concern over the possible effect of radiation from fallout on the population ..." (NCRP 1960).

On the basis of its review, the NCRP stated that:

" ... we recommend that the population permissible dose for man-made radiation be based on the average natural background level. Although it is not our responsibility to determine the exact level, we believe that the population permissible somatic dose for man-made radiations, excluding medical and dental sources, should not be larger than that due to natural background radiation, without a careful examination of the reasons for, and the expected benefits to society from a larger dose" (NCRP, 1960, page 485). Since the NCRP, in its report, assumed that the background dose rate was 100 mrem per year (NCRP, 1960, page 485), this statement implied that the dose rate limit that they were recommending for the general population was 100 mrem (0.1 rem) per year.

III. Relevant Standards and Guides—Federal Radiation Council

In 1959, the Congress (Public Law 86-373) established the Federal Radiation Council (FRC) to "... advise the President with respect to radiation matters, directly or indirectly affecting health, including guidance for all Federal agencies in the formulation of radiation standards and in the establishment and execution of programs of cooperation with States...". The formation of the FRC was an outgrowth of public hearings held in 1957 by the Joint Committee on Atomic Energy, on the "The Nature of Radioactive Fallout and Its Effects on Man." In all of these hearings, questions on the biological effects of radiation and protection against excessive exposure to radiation received attention.

1. FRC Report No. 1

In 1960, the FRC issued Report No. 1 (FRC 1960) in which they established the concept of a Radiation Protection Guide (RPG) which was defined as:

"... it is our basic recommendation that the yearly radiation exposure to the whole body of individuals in the general population (exclusive of natural background and the deliberate exposure of patients by practitioners of the healing arts) should not exceed 0.5 rem" (paragraph 5.3, page 26).

Because of continuing concern about the potential genetic effects, the FRC went on to state:

"When the size of the population group under consideration is sufficiently large, consideration should be given to the genetically significant population dose. The Federal Radiation Council endorses in principle the recommendations of such groups as the NAS-NRC, the NCRP, and the ICRP concerning population genetic dose, and recommends the use of the Radiation Protection Guide of 5 rem in 30 years (exclusive of natural background and the purposeful exposure of patients by practitioners of the healing arts) for limiting the average genetically significant exposure of the total U. S. population. The use of 0.17 rem per year ... is likely in the immediate future to assure that the gonadal exposure Guide is not exceeded" (paragraph 5.5, page 27).

2. FRC Report No. 2

In 1961, the FRC issued Report No. 2 (FRC 1961) in which they recommended:

"... RPG's for the thyroid gland of 1.5 rem per year for individuals and 0.5 rem per year to be applied to the average of suitable samples of an exposed group in the general population as representing a reasonable balance between biological risk and benefit to be derived from useful applications of radiation and atomic energy" (paragraph 2.9, page 11).

3. FRC Report No. 5

Confirming that their previous recommendations applied to "normal peacetime operations," in 1964 the FRC issued Report No. 5 (FRC 1964) in which they specifically addressed the concept of Protective Action Guides (PAGs) for application in the case of unusual concentrations of radionuclides in various foods due to the atmospheric testing of nuclear weapons (Introduction, page 1).

At that time, the FRC noted that a report of a panel of experts of the NAS-NRC Committees stated that:

> " ... although therapeutic doses from iodine-131 to the thyroid have been in the range of a few thousands rads upwards, iodine-131 has not been identified in a causative way with the development of thyroid cancer in humans, except in one doubtful case. X-ray doses to the thyroid appear to be from 5 to 15 times as effective in producing biological changes as iodine-131" (Section IV, page 12).

On the basis of this and other information, the FRC stated:

> "Considering existing information on the biological risks associated with doses from iodine-131 and the kinds of protective actions available to avert the dose from iodine-131 that has been deposited on pastures by dairy cows, the Council has concluded that such protective action as the diversion of milk or the substitution of stored feed for pasturage to avert individual doses less than 30 rads would not usually be justifiable under the conditions considered most likely to occur. This dose is recommended as the Protective Action Guide for iodine-131" (Section IV, page 12).

The FRC went on to say:

> " ... it is assumed that the majority of the individuals do not vary from the average by a factor greater than three. As an operational technique, it is considered that the PAG will not be exceeded if the average projected doses to the thyroids of a suitable sample of the population do not exceed 10 rads. A suitable sample is considered to consist of children of approximately one year of age using milk from a reasonably homogeneous supply" (Section IV, page 12). "The PAG is stated in terms of the projected dose; i.e., the dose that might otherwise be received if the protective action were not initiated" (Section IV, page 13).

In terms of the need for special consideration for children, the FRC stated that for I-131:

> "... a given intake would result in a ten times larger dose to the thyroid of a one year old child (thyroid weight 2 grams) than to an adult (thyroid weight 20 grams)" (Section IV, page 11).

In terms of operating criteria, the FRC stated:

> "A total intake of iodine-131 of 600 nanocuries would result in a dose to about 10 rads to a 2 gram thyroid" (Section IV, page 14).

IV. Current Standards and Guides—International Commission on Radiological Protection, National Council on Radiation Protection and Measurements, and the U. S. Nuclear Regulatory Commission

Although they would not have applied during the period from 1950 to 1970, it is useful also to consider the current recommendations and/or regulations of the International Commission on Radiological Protection (ICRP), the NCRP, and the U. S. Nuclear Regulatory Commission (U. S. NRC) relative to dose rate limits for members of the public.

1. ICRP Publication 60

In Publication 60, the ICRP (ICRP 1991) recommends that:

> "... the limit for public exposure should be expressed as an effective dose of 1 mSv in a year. However, in special circumstances, a higher effective dose could be allowed in a single year, provided that the average over 5 years does not exceed 1 mSv per year" (paragraph S40, page 75).

2. NCRP Report No. 116

In Report No. 116, the NCRP (NCRP 1993) recommends:

> "For continuous (or frequent) exposure, it is recommended that the annual effective dose not exceed 1 mSv." "Furthermore, a maximum annual effective dose limit of 5 mSv is recommended to provide for infrequent annual exposures. An annual effective dose limit recommendation of 5 mSv is made because annual exposures in excess of the 1 mSv recommendation, usually to a small group of people, need not be regarded as especially hazardous, provided it does not occur often to the same groups and that the average exposure to individuals in these groups does not exceed an average annual effective dose of about 1 mSv" (Section 15, page 46).

3. U. S. Nuclear Regulatory Commission

In its "Standards for Protection Against Radiation" (Title 10, Code of Federal Regulations, Part 20), issued in 1991, the U. S. NRC has stipulated:

> "Each licensee shall conduct operations so that — (1) The total effective dose equivalent to individual members of the public from the licensed operation does not exceed 0.1 rem (1 millisievert) in a year, exclusive of the dose contributions from background radiation, any medical administration the individual has received, voluntary participation in medical research programs, and the licensee's disposal of radioactive materials into sanitary sewerage ..." (paragraph 20.1301).

In summary, all three organizations recommend the restriction of average long-term effective (whole body) dose rates to the public to no more than 1 mSv (0.1 rem) per year.

4. U. S. Environmental Protection Agency and Food and Drug Administration

The U. S. Environmental Protection Agency and the Food and Drug Administration in their guides to emergency and protective action insofar as dose to the thyroid is concerned set the standards for action as 5-25 rad and 1.5-15 rad, respectively. Their ingestion protective action guides are set at 1.5 rad and 15 rad, the former being the preventive action guide and the latter the emergency action guide. At thyroid doses of 25 rad or greater administration of stable iodine is recommended.

References

FRC (Federal Radiation Council). 1960. Background Material for the Development of Radiation Protection Standards. Report No. 1. Washington, DC.

FRC (Federal Radiation Council). 1961. Background Material for the Development of Radiation Protection Standards. Report No. 2. Washington, DC.

FRC (Federal Radiation Council). 1964. Background Material for the Development of Radiation Protection Standards. Report No. 5. Washington DC.

ICRP (International Commission on Radiological Protection). 1991. 1990 recommendations of the International Commission on Radiological Protection. ICRP Publication 60, Annals of the ICRP 21(1-3). Oxford: Pergamon Press.

NCRP (National Council on Radiation Protection and Measurements). 1953. Maximum Permissible Amounts of Radioisotopes in the Human Body and Maximum Permissible Concentrations in Air and Water. Handbook 52. Washington, DC: National Bureau of Standards.

NCRP (National Council on Radiation Protection and Measurements). 1954. Permissible Dose from External Sources of Ionizing Radiation. Handbook 59. Washington, DC: National Bureau of Standards.

NCRP (National Council on Radiation Protection and Measurements). 1959. Maximum Permissible Body Burdens and Maximum Permissible Concentrations of Radionuclides in Air and Water for Occupational Exposure. Handbook 69. Washington, DC: National Bureau of Standards.

NCRP 1960. Somatic dose for the general population. Report of an Ad Hoc Committee. Science 131:482-486.

NCRP (National Council on Radiation Protection and Measurements). 1993. Limitation of Exposure to Ionizing Radiation. NCRP Report No. 116. Bethesda, MD: National Council on Radiation Protection and Measurements.

Appendix F

Screening for Thyroid Cancer: Background Paper

Karen Eden, Ph.D.
Mark Helfand, M.D.
Susan Mahon, M.P.H.

Between 1951 and 1962 the United States Atomic Energy Commission conducted more than 100 aboveground detonations of nuclear weapons at the Nevada Test Site. Parts of Idaho, Montana, Utah, and Colorado had the highest exposure, but radioactive fallout was deposited throughout the United States. Infants and small children who drank milk or ate fresh vegetables contaminated with fallout were exposed to the highest doses of radiation, and as adults these persons are most at risk for developing thyroid abnormalities.

In 1997, the National Cancer Institute (NCI) released a report estimating the amount of radiation exposure Americans received from the nuclear tests (USDHHS, 1997). The report states that the average dose of radiation received by each American was 0.02 gray (Gy). Average exposures in some Western and Midwestern states were higher as much as 0.16 Gy in some areas. In addition to place of residence, the source and quantity of milk consumed by infants and young children can be used to help predict the dosage of the radioactive iodine (I-131) reaching the thyroid gland.

The NCI findings raised concerns that individuals exposed to higher dosages of I-131 might have a high risk of developing thyroid cancer later in life and that early detection and intervention for thyroid cancer could be useful for those persons.

To justify screening, there must be evidence that screening tests can detect thyroid cancer accurately and safely in asymptomatic individuals, that early treatment decreases mortality or morbidity from thyroid cancer when compared with delaying treatment until symptoms occur, and that the benefits of screening clearly outweigh the adverse effects of the screening program itself. Unfortunately, evidence about the ability of early detection to prevent morbidity and mortality from thyroid cancer is incomplete. In contrast to the case of screening for breast cancer, for example, there are no randomized, controlled trials of screening for thyroid cancer, either in the general population or in high-risk groups. Instead, the effect of screening must be inferred from observational studies regarding the prevalence of undiagnosed thyroid cancer and the potential benefit of early detection.

In this paper we address whether physicians should screen for thyroid cancer seen in asymptomatic patients thought to have been exposed to I-131. We focus on studies of screening in the general population and in high-risk groups. We examine the accuracy of the tests used for screening, the number of cancer patients detected with screening, and evidence that treatment of cancers found by screening improves outcomes.

Definitions

Screening is "the application of a test to detect a potential disease or condition in a person who has no known signs or symptoms of that condition at the time the test is done" (Eddy, 1991). Studies of screening can be classified according to the setting in which the decision to screen takes place. In *clinic-based screening*, or *casefinding*, a screening test is performed in patients who visit a primary-care physician for an unrelated reason. Studies of casefinding programs provide the most realistic estimates of the effects and costs of screening in a clinic or office practice, but there have been very few studies of casefinding for thyroid cancer. *Population-based studies* contact, recruit, and follow patients in the context of an epidemiologic research effort. Such studies show the extent of unsuspected thyroid cancer in a population sample, but they do not reflect the yield or costs of screening in clinics or providers' offices. We used population-based studies as a benchmark against which the yield and benefits of clinic-based screening programs could be measured. Finally, studies of *monitoring in high-risk groups* describe efforts to monitor individuals with occupational exposure to radiation or with a history of head and neck irradiation for benign or malignant conditions. Such studies could be less relevant to screening in populations exposed to the much lower doses of radiation from the Nevada Test Site.

Background: Challenges in Developing and Assessing a Screening Proposal

Before the consequences of screening can be estimated, it is necessary to formulate the screening problem. To formulate a strategy researchers must specify the intended population for screening; the screening tests, follow-up tests, and treatments that will be used; and the type of outcomes influenced by the tests. In this section, we enumerate the information gaps, and we suggest a strategy for a "baseline" screening program.

What Is the Target Population? How Can It Be Reached?

For screening to be effective, there should be a substantial prevalence of undiagnosed disease in the target population. It is difficult to estimate this prevalence unless the target population for screening is clearly defined and is similar to populations used for published studies. Current age, place of residence as an infant or young child, and current residence might be used as criteria for screening.

One approach to identifying patients might be for primary-care providers (or public health agencies) throughout the United States to ask questions designed to identify exposed individuals. If this approach is taken, those questions are the initial test for screening, and their sensitivity and specificity for identifying high-risk individuals would need to be measured to estimate the prevalence of disease in the screened population. Another approach would be to target individuals between the ages 35 and 55 who are current residents of counties that had high exposure to radiation; this would omit individuals who had moved away and could include low risk individuals who had moved to those areas as adults. Estimating the effectiveness of this approach would require information on the rate of migration to and from these areas and areas that received lower doses. A third approach would replicate the efforts of epidemiologic studies to contact and track residents throughout the United States and have them report to a single, organized source of care. Finally, an effort to teach and promote thyroid self-examination could be the primary tool for screening. Studies of screening programs do provide information about screening in a defined geographic area, but little if any information is available about the efficacy of the other approaches.

Exposure from the Nevada Test Site detonations occurred more than 30 years ago. Exposed individuals could now be widely dispersed throughout the United States. People who have already been diagnosed and treated for thyroid cancer cannot benefit from screening, so the yield of screening will depend on the probability of having undiagnosed cancer many years after the exposure.

What Test Should Be Done? Who Should Do Them? How Should They Be Interpreted?

To estimate the effect of screening, the choice and interpretation of screening and follow-up tests must be specified. Either a physical examination or an ultrasound examination can be used to screen for thyroid cancer. Both tests identify thyroid nodules that could be cancerous. If a nodule is found, a fine-needle aspiration (FNA) is done to identify cell types found. Individuals found to have malignant cells, or cells that could be malignant, could undergo surgery and other treatments for thyroid cancer.

The interpretation and accuracy of the tests vary. Whether palpation or ultrasound is used, it is necessary to specify how an abnormal result is defined and what action will be taken if the result is normal or abnormal. One variable is the size of the nodule that prompts confirmation by FNA. In some centers most patients with microadenomas are followed with physical examinations (Tan and Gharib, 1997); in other programs, serial ultrasound test or no tests at all might be done. Recommendations for the management of cystic nodules and of small thyroid cancers and protocols for the management of FNA results also vary. No single protocol has been adopted for screening programs.

How the tests are performed also can affect results. For ultrasound, relevant characteristics might include the resolution of the image and the qualifications and experience of the personnel who perform and interpret the test. For physical examination, a relevant characteristic might be whether examiners were given special training in examining the thyroid and whether standardized measurements were taken as part of the examination. In general, population-based studies and studies of surveillance for thyroid cancer have used structured interviews and examinations by expert examiners, conditions that could be difficult to replicate throughout the nation. Proficiency is also an important influence on the accuracy and cost of confirmatory FNA (Hall et al., 1989).

Which Cancers Require Treatment? What Treatments Should Be Used?

To estimate the effectiveness of early treatment for thyroid cancer we must be able to predict what treatment will be done when a cancer or suspicious nodule is found. Recommendations for treatment of thyroid cancers vary among studies and centers. For example, to manage small, localized papillary carcinomas, which are common in individuals exposed to ionizing radiation, some experts prefer total thyroidectomy; others recommend lobectomy with or without adjuvant therapy with ablative doses of I-131. More aggressive therapies could reduce the recurrence rate but cause higher rates of complication. In a nationwide initiative, individual surgeons' recommendations also might vary, making it difficult to estimate the rate of recurrence or complications that would be associated with screening.

What Health Outcomes Should Be Considered? How Do Screening Outcomes Compare with Usual-Care Outcomes?

Health outcomes relevant to screening for thyroid cancer include mortality and morbidity related to thyroid cancer and to the tests and treatments used in the screening program. Many studies report the number of cases of cancer detected by screening, but few report observed health outcomes in the screened population, and none has made an unbiased comparison of the outcomes of screening with the outcomes of usual care. As a result, the reduction in morbidity and mortality associated with screening can only be inferred from indirect evidence.

Figure F.1 illustrates the type of evidence that could be used to make these inferences. Arcs 1 and 2 indicate direct evidence comparing the mortality and morbidity associated with screening against that of usual care; such evidence is

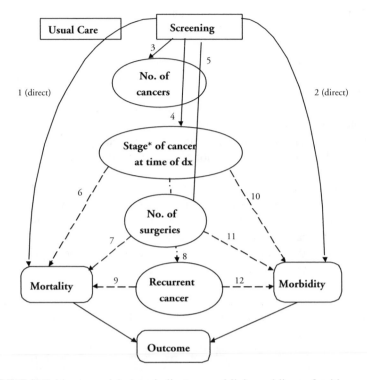

FIGURE F.1 Evidence model. Arcs indicate causal links and lines of evidence. Arcs 1 and 2 represent direct (experimental) evidence linking screening to morbidity and mortality. Arcs 3 to 5 represent evidence from observational studies of screening. Arcs 6-11 represent links for which the available evidence comes from epidemiologic and clinical observational studies not performed in the context of screening.

* "Stage" means any factor that can be used to predict outcome, including patient characteristics (e.g., age) and features of the cancer (size, histologic type, and presence of local or distant metastases).

not available. Studies of screening do provide evidence about the number of cancers found in a population (arc 3) and, to a lesser extent, the stages of these cancers at the time of detection (arc 4) and the number of surgical operations performed as a result of screening (arc 5).

Does finding these cancers and treating them earlier reduce mortality and morbidity? The stage of cancer at the time of diagnosis, cancer recurrence, and surgery performed are intermediate or surrogate outcomes of screening—they might be related to mortality and morbidity, but they are not a direct measurement of them. Because there is no direct evidence linking screening to health outcome, the effect of screening on mortality and morbidity is inferred. Arcs 6 through 12 illustrate causal links between the intermediate measures and health outcomes. Arc 6 is the relationship of cancer stage at the time of detection to mortality. Arc 7 is mortality due to fatal complications of surgery and anesthesia. Arc 8 is the relationship between stage of cancer at the time of detection and the likelihood of developing recurrent cancer. Arc 9 is the effect of recurrence on mortality. Similarly, morbidity depends on morbidity of cancer at the time of detection (arc 10), morbidity due to complications of surgery (arc 11), and morbidity due to recurrent cancer (arc 12). Arcs 10 and 12 are dashed because most studies report results for usual care rather than screening programs, and so they deal with populations that might not be similar to those included in screening programs.

Lack of information about current practice also makes it difficult to estimate the effectiveness of *not* screening; that is, of usual care. Screening identifies cancers that would be detected in usual care after some delay. How much is gained by finding these cancers earlier? In the absence of randomized trials, this benefit must be estimated from the length of the delay, the extent to which the cancer advances during the delay, and the effect of advancement on mortality and morbidity. Information about usual care is needed to estimate the length of delay and the rate of advancement during the delay, but such data are sparse. For this reason, the confidence range about any estimate of the benefit of early detection by screening will be wide.

No screening proposal can satisfy all of these concerns completely, and uncertainties do not mean we should not screen for thyroid cancer. Unless a specific proposal is formulated, however, the consequences of screening cannot be estimated.

Baseline Strategy and Assumption

We examined two alternatives to usual care: screening with palpation and screening with ultrasound examination. We assumed that screening would take place in the primary-care setting among patients seeing a provider for reasons other than thyroid disease. We also assumed that providers, health systems, and public health officers could reliably identify individuals exposed to a specified dose of radiation by asking questions about risk factors identified in the NCI study of the Nevada nuclear test.

Figure F.2 depicts the assumptions we made about how screening and follow-up tests would be interpreted and what actions would be taken on test results. For the "palpation" strategy, FNA would be performed if the examiner palpated a nodule. If the FNA result was "malignant" or "suspicious," the patient would undergo thyroidectomy. If the FNA result was "benign," palpation would be repeated every 6 months. If, over time, the nodule were to enlarge, a repeat FNA would be done. If the FNA result was "insufficient" or "nondiagnostic," FNA would be repeated, possibly under ultrasound guidance. If the FNA was persis-

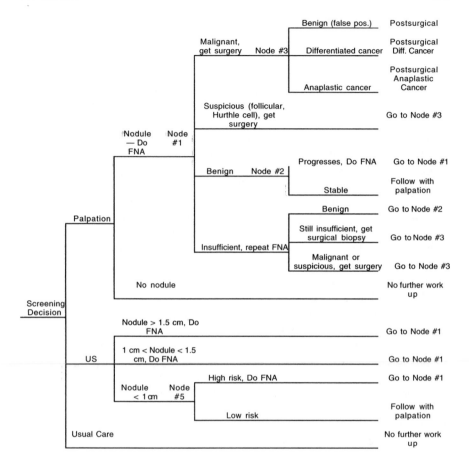

FIGURE F.2 Screening decision tree.

tently nondiagnostic, a lobectomy would be done to determine whether there was cancer.

These assumptions probably underestimate the use of ancillary tests in the follow-up of individuals found to have nodules. For example, in practice, serial ultrasound examinations, trials of levothyroxine, and scintigraphy are commonly used to evaluate or monitor patients with thyroid nodules, but we assumed that these interventions would not be part of a screening protocol and so did not attempt to assess the benefit or harm they produce.

In choosing between palpation and ultrasound as the first test for screening, an important issue is whether it is worthwhile to pursue small (<1 cm or <1.5 cm) impalpable nodules found by ultrasound. When palpation is the first test, this issue is moot, because by definition "incidentalomas" or occult thyroid cancers cannot be detected by palpation.

The "ultrasound" strategy is similar to the "palpation" strategy, except that the initial management of a nodule depends on its size. As shown in Figure F.2, if ultrasound is the first test, a decision about how to manage small nodules must be made (node 4). Several experts recommend against routine aspiration of all nodules, but their recommendations are intended for nodules found incidentally in individuals who do not exhibit risk factors for thyroid cancer. This recommendation is depicted as node 5 in Figure F.2, in which patients with additional risk factors proceed to aspiration, whereas "low risk" patients might be followed clinically. However, because individuals who have been exposed to radiation usually are considered "high risk," it is possible, in a program to screen individuals exposed to radiation, that aspiration with ultrasound guidance would be attempted in patients found to have occult nodules. If malignant or suspected cytology is found, surgical diagnosis and perhaps aggressive treatment are more likely— again because of the history of radiation exposure.

Thus, the foremost question in choosing between ultrasound and palpation is this: In patients with low levels of radiation exposure due to exposure to fallout from the Nevada experiments, do the potential benefits of detecting and treating occult thyroid cancers outweigh the potential harm posed by additional monitoring, surgery, and medical treatments that will result from screening? Because ambiguity about these decisions could reduce the benefits of screening, a screening recommendation about thyroid cancer should carefully specify what level of exposure makes a patient "high risk" with respect to the aggressiveness of follow-up of small nodules, what size nodules should prompt a work-up, and when aggressive surgical therapy should be applied.

Methods

We reviewed studies of screening for thyroid cancer and studies of the accuracy of tests commonly used to screen for thyroid cancer.

Data Sources

To find relevant articles on screening for thyroid cancer in high-risk populations, we searched the MEDLINE data base for papers published between 1966 and 1998 using the medical subject heading (MeSH) term *thyroid neoplasms* combined with the MeSH terms *mass screening, environmental exposure, radiation-induced neoplasms, power plants, nuclear reactors, radioactive fallout,* and *nuclear warfare,* and the text words *screen, screening, power plants, and nuclear reactors.* For studies of test performance, we searched the same data base, combining the MeSH terms *thyroid neoplasms* and *sensitivity and specificity.* For recent data on incidence and prevalence of thyroid cancer, we searched for papers published since 1987, and we restricted our focus to papers with the MeSH term *thyroid neoplasms* and the subheading *epidemiology*. These searches resulted in a total of 1,353 citations: 904 from the screening search, 249 from the test performance search, and 200 from the epidemiology search. To supplement our MEDLINE searches, reference lists from recent reviews were searched and articles recommended by thyroid cancer panel members were retrieved.

Papers were excluded unless they included information on at least one of the following areas:

- controlled studies
- screening in asymptomatic persons
- health outcomes
- screening by physical exam, ultrasound, or FNA
- recommendations for screening based on test results
- population-based studies
- size, stage, or type of nodule.

Studies of screening for congenital or familial thyroid disorders were also excluded.

Study Selection

In all, 56 studies were included in our review: Twenty-six addressed either screening in an exposed or unexposed population or surveillance in an exposed population; 23 addressed the accuracy of palpation, ultrasound, or FNA; and 10 addressed the incidence, natural history, treated history, or predictors of mortality in patients with thyroid nodules or thyroid cancer. (The numbers do not add to 56 because of overlap among paper topics.) We found only one study that attempted to examine whether a delay in the initial treatment of thyroid cancer was associated with decreased survival (Mazzaferri and Jhiang, 1994). We also included well-done review articles and meta-analyses about the epidemiology, course, and treatment of thyroid cancer and about the relationship between radiation exposure and the risk and type of thyroid cancer (Caruso and Mazzaferri, 1991; Fraker et al., 1997; Gharib, 1994; Gharib and Goellner, 1993; Giuffrida and Gharib, 1995; Moosa and Mazzaferri, 1997; Ron and Saftlas, 1996; Ron et al., 1995; Schlumberger, 1998; Tan and Gharib, 1997).

Data Extraction

We sought to answer several questions about four topics:

1. *Yield of screening.* How many individuals with thyroid nodules and cancer does screening identify?

2. *Accuracy of tests.* What are the sensitivity and specificity of the initial tests for screening (palpation or ultrasound) and of the confirmatory FNA test? What factors affect sensitivity and specificity?

3. *Consequences of a screening program.* Given the prevalence of disease and the accuracy of the diagnostic tests, how many individuals with thyroid cancer will screening detect? How many follow-up tests will be performed? How many healthy individuals will have false-positive test results that require additional evaluation? How many surgical operations will be done?

4. *Effectiveness.* Is there evidence that early treatment reduces the burden of illness? Specifically, how do the outcomes of screening compare with the outcomes of usual care? How might reducing the delay in diagnosis affect the potential complications of thyroid cancer?

From each study of screening or surveillance we extracted the following information: the setting in which screening or surveillance was performed, the risk status of the target population, the screening tests used, and the prevalence of thyroid nodules and of thyroid cancer found by screening. When it was available, we abstracted information on the characteristics of the cancers detected by screening (papillary, follicular, undifferentiated, localized, regional, distant metastasis, and so on.)

In studies that measured the performance of screening or confirmatory tests, we recorded the tests used, the gold standard determination of disease, the sensitivity and specificity of the tests, and the presence of biases (such as diagnostic-review bias) that could affect the reported results. We also abstracted the number of patients who underwent surgery to establish a final diagnosis.

Probability of Developing Thyroid Cancer

According to the American Cancer Society, 17,200 new cases of thyroid cancer will be diagnosed in 1998 in the United States (Landis et al., 1998). The incidence of thyroid cancer varies with sex and age. In the SEER registry, the average lifetime risk of being diagnosed with thyroid cancer was 0.27% for males and 0.66% for females (Ries et al., 1997). (For comparison, in a woman of "average risk," the risk of developing colon cancer or breast cancer is 6% or 9.4%, respectively.) In men, the incidence of thyroid cancer gradually increases until age 70-79, at which time the incidence is 8.6 per 100,000 (Ries et al., 1997). For women, the peak incidence, 13.2 per 100,000, is reached in the 50-54 year age group.

The probability of developing thyroid cancer depends on the presence of several risk factors. Exposure to ionizing radiation is the most well-established risk factor for thyroid cancer (McTiernan et al., 1984; Ron et al., 1995). In a population-based study done in Connecticut from 1978 to 1980, about 9% of thyroid cancers were attributable to prior irradiation of the head and neck (Ron et al., 1987). Most studies of this risk factor involve patients who had received external radiation to the head and neck for benign conditions (Pottern et al., 1990; Ron et al., 1988, 1989; Schneider et al., 1993; Shore et al., 1993; Tucker et al., 1991). In those studies, an increased incidence of thyroid cancer was observed among people who had received radiation doses as low as 10 rad (0.10 Gy) (Ron et al., 1988; Schneider et al., 1993), and the incidence rose linearly up to a dose of 1,000 rad (10 Gray) (Ron et al., 1989).

Mortality and Morbidity of Thyroid Cancer

In 1995 all-stage 5-year survival for thyroid cancer was 92.4% for males and 95.9% for females. In 1995 there were 1,120 deaths due to thyroid cancer in the United States (Wingo et al., 1995); 1,200 deaths are expected in 1998 (Landis et al., 1998). In 1994 the lifetime risk of dying from thyroid cancer was 0.04% for males and 0.07% for females. The risk of dying from thyroid cancer is much lower than the risk of developing thyroid cancer because most people with thyroid cancer, die of other causes.

The risk of death depends on age and on the characteristics of the thyroid cancer. Longitudinal studies of the treated history of thyroid cancer indicate that age at the time of diagnosis, tumor size, local invasion, tumor cell DNA content, and regional or distant metastasis are associated with mortality from thyroid cancer (Fraker et al., 1997). Because most deaths occur in elderly people who have undifferentiated, metastatic cancers, average rates overestimate mortality in younger and middle-aged individuals, who tend to have papillary and follicular cancers (Ron, 1996). Although distant metastases are present in only 5% of thyroid cancers at the time of diagnosis, patients in this group account for about 60% of deaths. At the other extreme, 56% of cancers are localized to the thyroid gland at the time of diagnosis, but only 4 patients per 1,000 in this group die from thyroid cancer within 5 years (USDHHS, 1997).

SCREENING TESTS

History and Physical Examination Palpation

A thorough clinical evaluation for thyroid disease usually includes ascertainment of age, gender, family history of thyroid disease, history of hormonal problems or prior neck irradiation, and visual inspection and palpation of the thyroid gland by a clinician (Ashcraft and Van Herle, 1981; Brander et al., 1992). The goal of palpation is to detect nodules and to assess the size of the thyroid gland. When a nodule is palpated, the examiner notes its firmness and attachment to underlying structures and attempts to determine whether a single nodule or multiple nodules are present. Additionally, nearby lymph glands are palpated to detect enlargement (Khafagi et al., 1988; Spiliotis et al., 1991). The examiner asks whether the patient has noticed it, how long it has been present, and whether the patient has had difficulty swallowing or has had hoarseness. Although the presence of these conditions increase the likelihood that a nodule is malignant (Holleman et al., 1995; Khafagi et al., 1988; Merchant et al., 1995; Spiliotis et al., 1991), it is usually not possible to tell when nodules are benign or malignant by palpation alone (Ashcraft and Van Herle, 1981).

Three studies of screening in asymptomatic individuals used palpation and ultrasound to detect thyroid nodules (Table F.1). These studies used ultrasound as the gold standard determination of the presence of a nodule; the sensitivity of palpation for all sized nodules was 0.10 to 0.31, which indicates that a high proportion of nodules detected by ultrasound were too small to be palpated (Ezzat et al., 1994; Inskip et al., 1997; Mettler et al., 1992). This means that for a person with a nodule of any size, there is a probabiltiy of 0.10 to 0.31 that it will be palpated. The specificity of palpation ranged from 0.95 to 1.0. This means that for a person without a nodule, there is a probability of 0.95 to 1.0 that no nodule will be palpated.

One study group consisted of 100 healthy individuals who had no history of exposure to radiation. Twenty-one nodules were palpated in these individuals; 47 patients had nodules 0.5 cm or larger in diameter as shown by ultrasound examination, and 20 had nodules smaller than 0.5 cm. The overall sensitivity of palpation was 31% (21/67). For nodules larger than 0.5 cm the sensitivity was 0.45 (21/47), and for nodules larger than 1 cm the sensitivity was 0.84 (21/25).

In subjects who have been exposed to radiation, the sensitivity of palpation to find nodules could be lower. In one study, 18 nodules were palpated in a group of 1,060 children and adults living near Chernobyl (Mettler et al., 1992); in the same population, ultrasound detected 75 nodules larger than 0.5 cm in diameter (22 nodules, 0.5-0.9 cm, 44 nodules, 1-2 cm, 9 nodules >2 cm). Only 8 of the 18 palpated nodules were confirmed with ultrasound; 10 of the palpable nodules were not seen on ultrasound and were classified as false-positive results. For nodules larger than 2 cm, the sensitivity of palpation was 0.89, but sensitivity for 1-2 cm nodules was only 0.38.

In another study, 1,979 Estonian Chernobyl cleanup workers were examined from 1-9 years after their exposure at the reactor site (Inskip et al., 1997). In all, nodules were detected by ultrasound; of these 175 were larger than 0.5 cm, 88 were larger than 1 cm, and 35 were larger than 1.5 cm. In a study of patients screened 22 years or more after childhood head and neck irradiation (Schneider et al., 1997), palpation identified only 5 of 11 nodules 1.5 cm or larger, while ultrasound found 54 more.

Because the decision to examine and study a nodule is affected by its size, the ability to estimate the size of a nodule can be important. Clinicians often underestimate the size of palpable nodules and cannot reliably determine whether nodules are single or multiple (Brander et al., 1992; Tan et al., 1995). Some studies suggest that palpation is sensitive for nodules larger than 1.5 cm, others confirm the poor results of the studies discussed above. In one study, for

TABLE F.1 Sensitivity and Specificity of Palpation to Detect Nodules (using ultrasound as the gold standard)

Author, Year	Study sample and setting	P (nodule)	Palpation by 2 clinicians?	US Resolution	Size criteria for nodule	Sensitivity	Specificity	LR+	LR-	PV+	PV-
Ezzat, 1994	Screening of 84 female and 16 male unexposed adult American volunteers (n = 100)	0.67	y	10 MHz	All Sizes	0.31	1.00	NA[a]	0.69	1.00	0.42
Inskip, 1997	Screening of 1984 exposed adult clean up workers from Chernobyl (n = 1984)	0.11	n	7.5 MHz	All Sizes (FNA if >1 cm)	0.22	0.95	4.097	0.83	0.32	0.91
Mettler, 1992	Screening of 1060 residents of villages around Chernobyl (exposed and not exposed). Ages 5, 10, 40, 60 years (n = 1060)	0.08	y	7.5 MHz	>.5 cm	0.11	0.99	10.507	0.90	0.44	0.94

Definitions: Sensitivity (also called true positive rate) is the "likelihood that a diseased patient has a positive test."
Specificity (also called true negative rate) is the "likelihood that a nondiseased patient has a normal test."
LR+ (positive likelihood ratio) = sensitivity/(false positive rate). For example, a positive likelihood ratio of 10 means that a positive test is 10 times more likely to come from a patient with disease as from a patient without disease.
LR- (negative likelihood ratio) = (false negative rate)/specificity. For example, a negative likelihood ratio of .10 means that a negative test is 1/10th as likely to come from a patient with disease than from a patient without disease.
PV+ (positive predictive value) is "the fraction of patients with a positive test result who have disease."
PV- (negative predictive value) is "the fraction of patients with a normal test result who do not have disease."
[a] In Ezzat, all palpated nodules were confirmed with ultrasound, so this ratio is infinity.
SOURCE: Adapted from *Medical Decision Making*, HD Sox, MA Blatt, MC Higgins, KI Marton, Butterworth-Heinemann, 1988.

example, the sensitivity of palpation was 0.50 (14/28) for nodules 1-2 cm in diameter and 0.58 (19/33) for nodules larger than 2 cm (Brander et al., 1992).

Whether a single nodule or multiple nodules are present also can affect decision-making about a treatment, especially in individuals exposed to radiation who often have nodules scattered throughout the gland. In one study in which 151 single nodules were palpated 48% (73) of the patients actually had multinodular glands (Tan et al., 1995). Another study found that 63% (20/32) of palpated single nodules were shown by ultrasound examination to be multinodular (Brander et al., 1992)

In two studies of exposed populations, the positive predictive values for a palpated nodule were 0.32 and 0.44 (Inskip et al., 1997; Mettler et al., 1992), respectively. In one study, for example, 139 patients had palpable nodules, but only 44 of these were confirmed by ultrasound (positive predictive value, 0.32). In the same studies, the negative predictive values were 0.92 and 0.94. Because the prior probability of having no nodule was approximately 90% in these studies, a negative test result increases only slightly the probability of having no nodule.

The sensitivity of palpation appears to be 0.31 or less when all sizes of nodules are considered. However, sensitivity increases as size increases, particularly beyond 2 cm. Screening for nodules by palpation appears to have reasonable specificity—above 95%. Nevertheless, fewer than half of individuals thought to have a nodule by palpation actually do have one. For purposes of detecting whether a nodule is solitary, palpation miscategorizes approximately 50% of patients.

Ultrasound

To perform an ultrasound test on the thyroid, the patient lies supine with the neck hyperextended (Tan et al., 1995). A radiologist typically performs the test using an image scanner with 5-10 MHz linear transducers (Brander et al., 1992; Ezzat et al., 1994; Hsiao and Chang, 1994; Tan et al., 1995). An ultrasound test usually reveals the following details about thyroid nodules: echogenicity (reduced, normal, increased), "geneity" (homo, hetero), cystic component (nil, < 50%, 50%, >50%), calcification (present, absent), halo appearance (present, absent, doubtful), and nodularity (solitary, multiple) (Watters et al., 1992; Ashcraft and Van Herle; 1981). The information that a nodule is cystic or solid can be useful clinically. Cystic nodules often are managed by draining with a fine needle, and further evaluation depends on whether the nodule recurs.

Ultrasound is considered to be as accurate as is histologic examination of thyroid tissue for determining whether a thyroid nodule is present. The most important feature of ultrasound is its ability to detect clinically impalpable nodules. This feature makes ultrasound a highly sensitive test for screening, because such nodules, called incidentalomas or occult nodules, can contain carcinoma ("occult carcinoma"). In choosing between ultrasound and palpation as the first test, the foremost question is whether it is worthwhile to diagnose occult carcinoma, or whether screening is beneficial primarily in patients with relatively large ($\geq$1.5 cm) palpable nodules.

Confirmatory Tests: Fine Needle Aspiration

When a thyroid nodule is detected, an FNA confirmatory test often is performed (Bouvet et al., 1992; Cusick et al., 1990; Hamburger and Hamburger, 1985). FNA is a simple, relatively low-risk procedure that provides a cytologic sample to detect cancer, but it should be performed only by physicians experienced with the procedure (Gharib, 1994). It causes minor, local short-term pain and occasionally a small hematoma (Gharib, 1994; Gharib and Goellner, 1993). The use of FNA has greatly reduced the need for surgical excision to diagnose the cause of thyroid nodules (Fraker et al., 1997).

Fine-needle aspiration is usually an outpatient procedure. The physician palpates the thyroid gland while the patient lies supine on a bed. The skin around the thyroid is cleaned with alcohol, and a disposable needle (21-25 gauge) with a 10 mL syringe is inserted into the nodule (Gharib, 1994; Cusick et al., 1990). Suction is applied and the aspirate is sprayed onto glass slides (Haas et al., 1993). To reduce the number of false-negative results, several (2-6) aspirations are usually performed (Gharib and Goellner, 1993). Smears are made using the Papanicolaou, May-Grunwald-Giemsa, and/or Romanowsky-based Liu methods (Cusick et al., 1990; Gharib, 1994; Lin et al., 1997). The slides are reviewed by a cytopathologist who classifies the cells into one of the following groups (Holleman et al., 1995):

• *Benign.* Aspirated material adequate and consists of thyroid cell groups without signs of malignancy or consists of cyst fluid.
• *Follicular.* Little or no colloid with microfollicular cell groups. Differentiation between follicular adenoma and follicular carcinoma not possible.
• *Suspicious.* Signs suggestive but not diagnostic of malignancy in the aspirated material.
• *Malignant.* Several clearly malignant characteristics found.
• *Inadequate. (or unsatisfactory).* No conclusions because the smear does not contain enough material (fewer than 6 cell groups) or because the smear contains normal thyroid cells, suggesting that the neoplasm was not sampled.

In practice, 70% (range, 53%-90%) of FNA specimens are classified as benign; 4% (1%-10%) as malignant; 10% (5%-23%) as suspicious, follicular, or nondiagnostic; and 17% (15%-20%) as an inadequate or unsatisfactory sample (Fraker et al., 1997).

The result of FNA, along with other clinical information, is used to decide whether to proceed with surgical excision. The frequency of surgery that can be expected in a screening program, and the ratio of cancers to surgery, are important measures of the accuracy of FNA. In studies from major centers, an average of 18% of patients having FNA go on to have surgery (Gharib and Goellner, 1993; Haas et al., 1993; Spiliotis et al., 1991). Of patients who have surgery, it is estimated that 15-32% have cancer (Gharib and Goellner, 1993; Haas et al., 1993). The cancer probability depends on the criteria used to act on FNA results. Some investigators report that only the "malignant" category prompted surgery, but in many studies all patients with a suspicious or follicular FNA result also are taken to surgery (Bouvet et al., 1992; Cusick et al., 1990; Holleman et al., 1995; Khafagi et al., 1988; Morayati and Freitas, 1991).

The rate of nondiagnostic FNA results affects the rate of surgery. Because some authors do not differentiate between "unsatisfactory" and "indeterminate," the term "nondiagnostic" is used here for discussion. In some series, surgery was performed for all nondiagnostic FNAs (Cusick et al., 1990; Piromalli et al., 1992). Others took only patients with nondiagnostic results to surgery if other clinical symptoms were present: hoarseness, rapid growth of a nodule, or fixation to an underlying structure (Holleman et al., 1995; Jones et al., 1990; Khafagi et al., 1988; Lin et al., 1997; Merchant et al., 1995; Spiliotis et al., 1991).

The rate of nondiagnostic samples in referral clinics and hospitals ranges from 1% to 19% (Bouvet et al., 1992; Cusick et al., 1990; Haas et al., 1993; Holleman et al., 1995; Khafagi et al., 1988; Lin et al., 1997; Merchant et al., 1995; Ng et al., 1990; Ongphiphadhanakul et al., 1992; Piromalli et al., 1992; Spiliotis et al., 1991; Takashima et al., 1994). Studies that included a bedside evaluation of the sample with repeated aspiration as needed reported lower rates of nondiagnostic samples (Lin et al., 1997; Takashima et al., 1994). In one study, the researchers reported a nondiagnostic rate of 23% before implementing a bedside evaluation strategy to immediately repeat the test when nondiagnostic results were obtained (Khafagi et al., 1988). Those authors report dropping the nondiagnostic result rate to 12% after the bedside evaluation was in place. One group of investigators (Holleman et al., 1995) suggests that repeating all nondiagnostic FNAs and having two pathologists discuss the results would increase the number of diagnostic results. The probability of a nondiagnostic FNA result diminishes as operators become more experienced with the technique (Haas et al., 1993; Jones et al., 1990; Merchant et al., 1995; Piromalli et al., 1992) demonstrating the importance of having experienced physicians perform the technique.

The probability of nondiagnostic FNA results also is related to the size of the nodule being aspirated. One study found that 27% of the nodules smaller than 1.5 cm were nondiagnostic (Khafagi et al., 1988). In the same study, 16% of the nodules 1.5-3 cm and 13% of nodules larger than 3 cm were nondiagnostic.

Published studies of the accuracy of FNA use the findings from surgery as the gold standard determination of disease. These studies extended from the calculation of sensitivity and specificity patients who had benign FNA results and did not go to surgery. As Gharib and Goellner (1993) note, because of this diagnostic-review bias, published studies overestimate the sensitivity of FNA for thyroid cancer. Exclusion of those patients also causes FNA specificity to be underestimated.

In published studies, the reported sensitivity FNA for thyroid cancer range from 70% to 97% (Bouvet et al., 1992; Cusick et al., 1990; Haas et al., 1993; Hamburger and Hamburger, 1985; Holleman et al., 1995; Jones et al., 1990; Khafagi et al., 1988; Lin et al., 1997; Merchant et al., 1995; Morayati and Freitas, 1991; Ongphiphadhanakul et al., 1992; Piromalli et al., 1992; Spiliotis et al., 1991; Takashima et al., 1994; Watters et al., 1992). Specificity ranged from 52% to 98% for the same studies. The apparent sensitivity and specificity depend on how the investigators classify nondiagnostic FNA results. Examination of sensitivity and specificity results revealed that the surgical decisions related to nondiagnostic FNA results had a strong relationship with sensitivity and an even stronger one with specificity.

To examine this relationship in more detail, studies that document whether patients with nondiagnostic FNA results go to surgery and also that list the histology reports were reanalyzed to report sensitivity and specificity with and without including the nondiagnostic results. A nondiagnostic FNA was classified as "positive" if the patient went to surgery. The results of this analysis are summarized in Tables F.2a and F.2b.

The analysis reveals that sensitivity increases slightly with the inclusion of nondiagnostic FNA results. Most of the patients with nondiagnostic results have benign follicular adenomas, and specificity drops by 4-20 percentage points. This suggests that quite a few more patients with benign nodules went to surgery than represented by specificity results, which typically exclude nondiagnostic FNA results.

In our analysis, the positive predictive value of FNA is the proportion of all surgical patients who prove to have thyroid cancer. Among the studies summarized in Table F.2b, the positive predictive values ranged from 0.28 to 0.71. Figure F.3 shows that the positive predictive value of disease decreased as the prevalence of disease decreased. The patients in these surgical series had clinically apparent signs or symptoms that brought them to medical attention in the usual course of care. Patients identified in a screening program might have a lower probability of disease, so the positive predictive value of FNA could be lower than that shown by Table F.2b and Figure F.3.

Because some patients with benign nodules develop cancer years later, most clinicians follow those patients with periodic physical examinations. Longitudinal studies of patients who had a benign FNA and did not have surgery demonstrate that a benign FNA is a reliable finding. In four studies, patients with untreated, cytologically diagnosed benign thyroid nodules were followed for several years for evidence of thyroid cancer (Jones et al., 1990; Kuma et al., 1994;

TABLE F.2a Sensitivity and Specificity of FNA

Author, Year	Study sample and setting	Total FNA	Surg/ FNA	Repeat FNA?	US-guided FNA?	Surgery of Benign FNA	Nondiag-nostic	Nondiagnostic Surgery
		Patient n						
Cusick, 1990	395 Scottish patients with thyroid nodules at regional thyroid clinic.	395	77.7%	y	?	57.80%	6.1%	100.0%
Holleman, 1995	112 Dutch outpatients referred by general practictioner for thyroid nodules.	112	47.3%	y	y	26.10%	10.7%	41.7%
Jones, 1990	175 British euthyroid outpatients.	175	46.9%	?	?		13.7%	70.8%
Khafagi, 1988	618 Australian euthyroid outpatients with thyroid nodules referred to thyroid clinic.	618	41.7%	n, later y	n	38.6%	33.0%	31.4%
Lin, 1997	3657 Taiwanese outpatients who were referred for palpable thyroid nodules.	3657	10.3%	y	y	7.0%	3.2%	48.6%
Merchant, 1995	111 British referred for thyroid nodules.	111	50.5%	y	?	43.4%	13.5%	46.7%
Piromalli, 1992	795 Italian outpatients referred for palpable thyroid nodules.	795	27.2%	y	?		5.3%	100.0%
Spiliotis, 1991	1200 Greek outpatients referred for thyroid disease.	1200	16.8%	?	?	13.2%	13.3%	6.3%

Definitions: Sensitivity (also called true positive rate) is the "likelihood that a diseased patient has a positive test."
Specificity (also called true negative rate) is the "likelihood that a nondiseased patient has a normal test."
LR+ (positive likelihood ratio) = sensitivity/(false positive rate). For example, a positive likelihood ratio of 10 means that a positive test is 10 times more likely to come from a patient with disease as from a patient without disease.
LR- (negative likelihood ratio) = (false negative rate)/specificity. For example, a negative likelihood ratio of .10 means that a negative test is 1/10th as likely to come from a patient with disease than from a patient without disease.
PV+ (positive predictive value) is "the fraction of patients with a positive test result who have disease."
PV- (negative predictive value) is "the fraction of patients with a normal test result who do not have disease."
SOURCE: Adapted from Medical Decision Making, HD Sox, MA Blatt, MC Higgins, KI Marton, Butterworth-Heinemann, 1988.

TABLE F.2b Sensitivity and Specificity of FNA

Author, Year	Test for Malignant, Suspicious, Follicular							Includes All Nondiagnostic FNAs as + tests — Test for Malignant, Suspicious, Follicular, Nondiagnostic						
	Sensitivity	Specificity	LR+	LR-	p (cancer nodule)	PPV	NPV	sens. w/ nondiag.	spec. w/ nondiag.	LR+	LR-	p (cancer nodule)	PPV	NPV
Cusick, 1990	0.76	0.58	1.84	0.40	0.58	0.720	0.64	0.78	0.55	1.73	0.39	0.58	0.71	0.64
Holleman, 1995	0.84	0.52	1.74	0.31	0.40	0.533	0.83	0.85	0.45	1.56	0.33	0.40	0.49	0.83
Jones, 1990	0.65	0.69	2.13	0.50	0.40	0.585	0.75	0.70	0.52	1.46	0.58	0.40	0.46	0.75
Khafagi, 1988	0.89	0.72	3.17	0.15	0.20	0.436	0.97	0.91	0.52	1.91	0.17	0.20	0.28	0.97
Lin, 1997	0.80	0.99	59.58	0.20	0.31	0.963	0.92	0.82	0.82	4.58	0.22	0.31	0.65	0.92
Merchant, 1995	0.71	0.91	8.33	0.31	0.29	0.769	0.89	0.73	0.78	3.34	0.34	0.29	0.55	0.89
Piromalli, 1992	0.95	0.84	5.83	0.06	0.33	0.740	0.97	0.95	0.65	2.72	0.07	0.33	0.54	0.97
Spiliotis, 1991	0.89	0.86	6.61	0.12	0.16	0.556	0.98	0.90	0.75	3.66	0.13	0.16	0.40	0.98

Definitions: Sensitivity (also called true positive rate) is the "likelihood that a diseased patient has a positive test."
Specificity (also called true negative rate) is the "likelihood that a nondiseased patient has a normal test."
LR+ (positive likelihood ratio) = sensitivity/(false positive rate). For example, a positive likelihood ratio of 10 means that a positive test is 10 times more likely to come from a patient with disease as from a patient without disease.
LR- (negative likelihood ratio) = (false negative rate)/specificity. For example, a negative likelihood ratio of .10 means that a negative test is 1/10th as likely to come from a patient with disease than from a patient without disease.
PV+ (positive predictive value) is "the fraction of patients with a positive test result who have disease."
PV- (negative predictive value) is "the fraction of patients with a normal test result who do not have disease."
SOURCE: Adapted from Medical Decision Making, HD Sox, MA Blatt, MC Higgins, KI Marton, Butterworth-Heinemann, 1988.

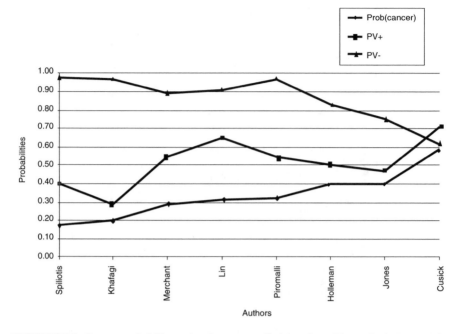

FIGURE F.3 Cancer probability as it relates to predictive values. Notes: Prob (cancer) is probability of thyroid cancer; PV+, positive predictive value, is the fraction of patients with a positive test who have thyroid cancer; PV–, negative predictive value, is the fraction of patients with a normal test who do not have thyroid cancer.

Morayati and Freitas, 1991; Piromalli et al., 1992; Tan and Gharib, 1997). In one study (Jones et al., 1990), 93 patients with benign FNA results were followed for at least 2 years with annual ultrasound and semiannual physical exams. There were no thyroid malignancies found in the follow-up period. In another study, 537 patients with benign FNA results were followed at 6-month intervals for 2-10 years (Piromalli et al., 1992). In this cohort no malignancies were found. An additional 42 patients in this group were lost to follow-up. In a follow-up study of 576 patients with benign FNA results and no surgical treatment (Morayati and Freitas, 1991), there were no documented thyroid malignancies after an average of 4.2 years of follow-up. In a long-term follow-up study by Kuma et al. (1994), cited in Tan and Gharib (1997), only 1 patient of 134 with nodules cytologically diagnosed as benign was diagnosed with a papillary carcinoma during a follow-up period of 9-11 years.

Consequences of Screening

We reviewed studies of screening programs to estimate the likely consequences of a screening program in individuals exposed to fallout from the Nevada tests. For the screened population, we estimated the probability that a screened individual would be found to have a nodule; the probability that an individual would be found to have thyroid cancer, the distribution of size, histologic type, and spread of cancers; and the probability that a screened individual would undergo surgery. We found no evidence from studies of screening that we could use to estimate other important events related to health outcomes, such as the probability of long-term survival in patients found to have cancer or the rates of surgical complication. As mentioned earlier, these events must be estimated indirectly from studies performed in other settings.

Studies

Of the 26 studies of screening we selected, 12 concerned healthy adults with no history of exposure to radiation. Of these, 4 were population-based studies of healthy adults with no prior exposure to radiation (Brander et al., 1991; Tunbridge et al., 1977; Vander et al., 1968), 4 were studies conducted at a single institution involving unexposed volunteers (Ezzat et al., 1994; Hsiao and Chang, 1994), 1 was a mass screening of healthy adults (Ishida et al., 1988), and 3 were screening programs of healthy adults using casefinding (Bruneton et al., 1994; Miki et al., 1993; Woestyn et al., 1985).

Eight studies concerned individuals exposed to fallout from nuclear detonations. Three studies examined thyroid disease in workers after the Chernobyl nuclear reactor accident (Inskip et al., 1997; Ivanov et al., 1997) and in Chernobyl residents (Mettler et al., 1992). Three papers looked at residents of the Marshall Islands exposed to fallout from hydrogen bomb testing (Hamilton et al., 1987; Howard et al., 1997; Takahashi et al., 1997); 1 examined thyroid disease in survivors of the 1954 atomic bomb blast in Nagasaki (Nagataki et al., 1994); and 1 looked at residents of Utah, Nevada, and Arizona who were potentially exposed to fallout from nuclear weapons testing (Kerber et al., 1993). Time since exposure ranged from 1 year in Chernobyl to more than 40 years.

Six studies concerned surveillance for thyroid cancer in individuals exposed to radiation from medical interventions. Five papers described recall studies of adults who were exposed as children to medical radiation, either for benign conditions (Favus et al., 1976; Ron et al., 1984; Royce et al., 1979; Shimaoka et al., 1982) or for childhood cancer (Crom et al., 1997). One additional study concerned occupationally exposed hospital workers (Antonelli et al., 1995).

Tables F.3a and Table F.3b summarize results of the studies in 4 areas: probability of finding a nodule by palpation or ultrasound, probability of finding cancer, the proportion of nodules found to be cancerous, and the number of surgical operations performed as a result of screening. (Table F.3a provides descriptive data on the studies revealed and Table F.3b presents data probabilities calculated from the same studies).

Probability of Diagnosing a Nodule

The number of nodules found in screening programs depends on the population screened (unexposed or exposed), the test used (palpation or ultrasound), and the threshold size used to define a nodule (0.2 cm to 1.5 cm in various studies).

Unexposed Populations

Studies of screening unexposed adults with palpation find probabilities of nodules in the range of 0.01-0.05 (Brander et al., 1991; Ishida et al., 1988; Miki et al., 1993; Tunbridge et al., 1977; Vander et al., 1968). In unexposed populations, palpable thyroid nodules are more common in women, and their likelihood increases with age. In a population-based study in Whickham County, England (Tunbridge et al., 1977), 5.3% of women had palpable thyroid nodules. The probability for 18-24 year-old women was 4.7% and increased steadily to 9.1% for women over age 75. In the same study, only 0.8% of men had palpable thyroid nodules. In the United States, the probability of thyroid nodules for subjects enrolled in the Framingham study was similar (Vander et al., 1968): The all-age probability of palpated thyroid nodules among adult males was 1.5%; it was 6.4% among adult females.

More nodules will be diagnosed if ultrasound is used as the first test for screening. In studies performed in the general population, rates of 0.20-0.27 were observed (Brander et al., 1991; Miki et al., 1993). Women and older persons had higher rates. High-resolution ultrasound screening of 101 asymptomatic Finnish women (aged 49-58) detected nodules in 31 women (Brander et al., 1989). In France, 645 asymptomatic women and 355 asymptomatic men were screened using ultrasound (Bruneton, 1994). Forty-four percent of the women had thyroid nodules; 17.7% of the men had them. The probability of a thyroid nodule was 0.25 for participants under 50 years and increased to 0.42 for persons over age 50.

Exposed Populations

People who have been exposed to environmental and medical radiation have a higher prevalence of thyroid nodules than is found among the general population. Six population-based studies compared the number of nodules found through screening in radiation-exposed populations and in comparison groups (Hamilton et al., 1987; Howard et al., 1997; Kerber et al., 1993; Mettler et al., 1992; Nagataki et al., 1994; Takahashi et al., 1997). People exposed to the highest doses of radiation from the Nevada nuclear tests had a probability of 0.02-0.04 as determined by palpation; those exposed to lower doses had a probability of 0.02. In Marshall Islands tests, the probability in exposed groups was 0.07 (0.02-0.39) compared with 0.02 in people living farthest from the test site (Hamilton et al., 1987). Howard and co-workers (1997) found an even greater difference: 0.23 in exposed Marshall Islanders compared with 0.06 in unexposed

TABLE F.3a Studies of Screening for Thyroid Neoplasms

Author, year	Study sample and setting	Exposure	Time since exposure	Tests performed	Definition of a nodule	Group
Population-based studies						
Brander 1991	253 randomly selected adults in Finnish town	Unexposed	—	Questionnaire, PE, US	> 0.5 cm	
Brander 1989	101 women in a Finnish town, 1988	Unexposed	—		> 0.5 cm	
Ishida 1988	Mass screening: 152,651 Japanese women	Unexposed	—			
Tunbridge 1977	2,779 adults in an English town	Unexposed	—			
Vander 1954, 1968	5,108 adults in Framingham, MA without previous thyroid nodules	Unexposed	—			
Kerber 1993	2,473 adults in Nevada, Utah, Arizona	Fallout from nuclear weapons testing	7 to 35 years	Interview, PE; FNA & ^{123}I scan if palpable nodule	palpable	Dose = 50- >400 mGy (n=1418) Dose = 0-49 mGy (n=1055)
Hamilton, 1987	2,273 residents of 14 atolls of Marshall Islands	Fallout from BRAVO atomic bomb test on Bikini Island	33 yrs	PE	>1 cm	Most exposed islands (n = 1903) Least exposed (2 furthest atolls; n = 370)
Howard, 1997	Marshall Islands residents PE: 253 exposed, 227 comparison US: 177 exposed, 47 comparison	Fallout from nuclear weapons testing	PE: 1-40 yrs; US @ 40 yrs	PE (1954 to 1994) plus ultrasound (1994)	1 cm or 0.5 cm plus features of malignancy	Exposed (PE n = 253; US n=117) Unexposed (PE n = 227; US n = 47)
Takahashi, 1997	Residents of Marshall Islands	Fallout from nuclear weapons testing	42 yrs	PE, US		BRAVO cohort (n = 815) EOT cohort (n = 1062) AT cohort (n = 216)

TABLE F.3a continued

Author, year	Study sample and setting	Exposure	Time since exposure	Tests performed	Definition of a nodule	Group
Mettler, 1992	1,061 adults and children, population sample, Belarus	Chernobyl	4.5 yrs	PE, US	PE: dx of 2 MDs. US: >0.5 cm	Exposed (n = 597) Unexposed (n = 481)
Ivanov, 1997	167,862 Chernobyl cleanup workers	Chernobyl	1 to 8 yrs		—	1986 workers (mean dose 0.17 Gy; n = 77,663) 1987 workers (mean dose 0.10 Gy; n = 58,694) 1988-90 workers (mean dose 0.04 Gy; n = 31,565)
Inskip, 1997	1,984 Chernobyl cleanup workers	Chernobyl	4 to 9 yrs	PE, US, +FNA if >1 cm		
Nagataki, 1994	1,978 adult residents of Nagasaki, Japan	Atomic bomb	40 yrs		> 0.5 cm	Exposed (n = 1,043) Unexposed (n = 935)
Case finding studies						
Bruneton, 1994	1,000 healthy adult volunteers in France, 57% over 50 years of age, most of whom had an ENT infection within 3 months	Unexposed	—	US	> 0.3 cm	
Woestyn, 1985	300 healthy adults in Belgium referred for US of abdomen	Unexposed	—	PE, US	Prob > 0.3 cm	
Ezzat, 1994	100 healthy adult volunteers in California	Unexposed	—	PE, lab, US	FNA done if > 1 cm	
Hsaio, 1994	200 adults undergoing routine PE in Taiwan	Unexposed	—	PE, US	> 0.2 cm	

continues

Reference	Population	Exposure	Duration/age	Method	Criteria	n
Horlocker, 1986	1,000 patients being evaluated for suspected hyperparathyroidism in Rochester, Minnesota	Unexposed (46 had had head or neck irradiation)	—			
Miki, 1993	451 Japanese adults at employee physical or breast CA screening	Unexposed	—	PE, US		
Carroll, 1982	67 individuals undergoing carotid US	Unexposed		US, palpation if abnormal US		
Clinic-based monitoring in exposed groups						
Ron, 1984	443 adult volunteers in Israel exposed to 9 rads, mostly for benign conditions	Head and neck irradiation	About 30 yrs	PE, 99mTc scan if PE positive	Palpable	
Royce, 1979	217 exposed adults and 243 controls in Pennsylvania	Head and neck irradiation	9 to 42 yrs	Questionnaire, PE by 2 MDs, 99mTc if PE pos, surgery if + scan or nodule >1.5 cm	Palpable	Exposed (n = 214) Unexposed (n = 243)
Favus, 1976	1,056 adults in Illinois	Head and neck irradiation	12 to 58 yrs	Questionnaire, PE, 99mTc scan	Pos PE or 99mTc scan	
Shimaoka, 1989	1,500 adults in New York	Head and neck irradiation	not given; in 1979 mean = 25.6 yrs			
Crom, 1997	96 adult volunteers in Tennessee	Radiation therapy for childhood cancer (2-60 Gy)	5.6 to 22.8 yrs (median = 10.6)	Questionnaire, PE, lab studies, and US		
Antonelli, 1995	50 exposed Italian male workers compared to 100 unexposed hospital workers and 100 unexposed shipyard workers	Occupational exposure to x-rays in hospital workers	Duration of exposure: at least 10 yrs (mean = 22 yrs)	Questionnaire, PE, US	> 0.5 cm (by 2 operators); ^{131}I scan & FNA if > 1 cm	Exposed (n = 50) Unexposed (n = 200)

TABLE F.3b Studies of Screening for Thyroid Neoplasms

Author, year	Definition of a nodule	Group	Palpation				Ultrasound				
			Nodule P (range)	Cancer P (range)	Surgery n	P(Ca/Nodule)	Resolution MHz	Nodule P	Cancer P	Surgery n	P(Ca/Nodule)
Population-based studies											
Brander 1991	> 0.5 cm		0.05	0.000		0.00	7.5	0.27	0.000	0.004	0.00
Brander, 1989	> 0.5 cm						7.5	0.17			
Ishida 1988			0.04	0.001		0.03					
Tunbridge 1977			0.03								
Vander 1954, 1968			0.04	0.000	0.009	0.00					
Kerber 1993	palpable	Dose = 50–>400 mGy (n = 1418)	0.03 (0.02-0.04)	0.003 (0.000-0.008)		0.10					
		Dose = 0–49 mGy (n = 1055)	0.02	0.004		(0.00-0.27)					
Hamilton, 1987	>1 cm	Most exposed islands (n = 1903)	0.07 (0.02-0.39); solitary 0.03		—						
		Least exposed (2 furthest atolls; n = 370)	0.02		—						
Howard, 1997	1 cm or 0.5 cm plus features of malignancy	Exposed (PE n = 253; US n = 117)	0.23	0.07	0.135	0.30	7.5	0.25		0.014	
		Unexposed (PE n = 227; US n = 47)	0.06	0.02	0.033	0.33		0.26			
Takahashi, 1997		BRAVO cohort (n = 815)	0.16	0.02	0.016	0.06	7.5	0.33			
		EOT cohort (n = 1062)	0.14	0.02	0.023	0.07		0.30			
		AT cohort (n = 216)	0.042	0.02		0.17		0.12			

Study	Criterion	Group									
Mettler, 1992	PE: dx of 2 MDs.	Exposed (n = 597)	0.031		—	—	7.5	0.08			
		Unexposed (n = 481)	0.02					0.08			
Ivanov, 1997	—	1986 workers (mean dose 0.17 Gy; n = 77,663)		0.000	0.000	—					
		1987 workers (mean dose 0.10 Gy; n = 58,694)		0.000					0.001		
		1988-90 workers (mean dose 0.04 Gy; n = 31,565)		0.000							
Inskip, 1997			0.02		0.005		7.5	0.11	0.001		0.01
Nagataki, 1994	> 0.5 cm	Exposed (n = 1,043)					7.5	0.10	0.008	0.008	0.08
		Unexposed (n = 935)						0.05	0.001		0.02
Case finding studies											
Bruneton, 1994	> 0.3 cm						13	0.35			
Woestyn, 1985	Prob > 0.3 cm						5.5	0.19	0.000		
Ezzat, 1994	FNA done if > 1 cm		0.21 (0.09 solitary)				10.5	0.67[a]			
Hsaio, 1994	> 0.2 cm		0.08	0.009	0.689	0.113	7.5	0.16			
Horlocker, 1986			0.013				10	0.41	0.013	0.689	0.032
Miki, 1993			0.06				10	0.20	0.002	0.002	0.01
Carroll, 1982							7.5	0.13			
Clinic-based monitoring in exposed groups											
Ron, 1984	Palpable	Exposed (n = 214)	0.05	0.000	0.019	0.00					
Royce, 1979	Palpable		0.06	0.004	0.028	0.07					
		Unexposed (n = 243)	0.05	0.008		0.16					
Favus, 1976	Pos PE or 99mTc scan		0.13 (0.08 single)	0.057	0.256	0.38					
Shimaoka, 1989			0.34	0.01	0.057	0.03					

continues

TABLE F.3b continued

			Palpation				Ultrasound				
Author, year	Definition of a nodule	Group	Nodule p (range)	Cancer p (range)	Surgery n	P(Ca/ Nodule)	Resolution MHz	Nodule p	Cancer p	Surgery n	P(Ca/ Nodule) p
Crom, 1997			0.14				5 or 7	0.23	0.01		0.04
Antonelli, 1995	> 0.5 cm (by 2 operators); ^{131}I scan & FNA if > 1 cm	Exposed (n = 50)					7.5	0.38	0.00		0.00
		Unexposed (n = 200)						0.16	0.00		0.00

[a] Of which 0.14 are solitary nodules >0.5 cm.

groups. Mettler and co-workers (1992) found no difference in the probabilities of palpable or ultrasound-detected nodules in exposed versus unexposed population groups when highly contaminated villages near Chernobyl were compared with presumably unexposed control villages 4.5 years after the accident. In survivors of the Nagasaki atomic bomb blast, probabilities by ultrasound screening were 0.05 in unexposed groups and 0.10 in an exposed group (Nagataki et al., 1994).

Irradiation for benign medical conditions has been shown to result in an increased probability of thyroid nodules. Surveillance studies of people irradiated as children for benign conditions have found probabilities of 0.05-0.34 by palpation and 0.23-0.38 by ultrasound detection (Favus et al., 1976; Ron et al., 1984; Royce et al., 1979; Shimaoka et al., 1982). Antonelli and co-workers (1995) found a higher prevalence of nodules in hospital workers exposed to long-term, low-dosage x rays than was found in two unexposed control groups.

Nodule Size Distribution

Table F.4 shows the distribution of nodule sizes found in screening studies of exposed individuals. The information in the table can be used to estimate the additional number of nodules requiring follow-up when smaller nodules are considered to be significant.

Probability of a Thyroid Cancer Diagnosis

The yield of thyroid cancer in a screening program depends on the same factors that influence the probability of finding nodules. Moreover, the sensitivity of FNA and other test used to select patients for surgery can affect the prob-

TABLE F.4 Distribution of Nodule Size in Screening Studies of Exposed Patients[a]

Study	Total Nodules	<0.5 cm	0.5-0.99 cm	1.0-1.49 cm	≥1.5 cm
Inskip, 1997	201	12.9%	43.3%	26.4%	17.4%
Howard, 1997	117	NA	89.7%	10.3%	
Mettler, 1992	81	30.0 %		70.0%	
Shimaoka, 1982	19	NA	47.4%[b]	52.6%[c]	
Schneider, 1997	157	74.5%		25.5%	

NA Not available.

[a] All studies used ultrasound except Shimaoka.

[b] 0.5-1.0 cm.

[c] >1.0 cm.

ability of finding thyroid cancer. In general, more cancers are found in radiation-exposed subjects than in the general population and in subjects screened with an ultrasound test versus those screened by palpation. The decision to evaluate small nodules also will increase the yield of cancer diagnosis in screening programs. When higher resolution ultrasound is used, still more nodules and cancers will be detected.

Unexposed Populations

In the general population, 1-5% of nodules are found to be carcinomas; the probability that a nodule is cancerous appears to be higher for larger nodules detected by palpation than for smaller nodules found by ultrasound (Ishida et al., 1988; Miki et al., 1993).

Three population-based studies examined the rates of cancer found in unexposed people (Brander et al., 1991; Ishida et al., 1988; Miki et al., 1993). In a population of healthy Japanese women, 1 per 1,000 screened by palpation had thyroid cancer (Ishida et al., 1988); in a separate study of healthy Japanese adults who underwent an ultrasound examination at a routine physical examination or at a mammography screening center, 2 per 1,000 had thyroid cancer (Miki et al., 1993). In the Framingham study, no cancers at all were detected by palpation in a population sample of 5,108 American adults (Vander et al., 1968). Screening in 253 healthy Finns by palpation and by ultrasound revealed no cancers (Brander et al., 1989).

Exposed Populations

In comparison studies, more cancers are found by palpation in exposed than in unexposed populations. The Nevada Test Site study found probabilities of cancer of 0.008 in the most-exposed group and 0.004 in the least-exposed group (Kerber et al., 1993). Howard and co-workers (1997) found more cancers in exposed Marshall Islanders than they found in a comparison group (0.07 vs. 0.02), and Takahashi and co-workers (1997) found no difference in probability in exposed and unexposed groups—both probabilities were 0.02. These results involve palpation screening only, and in the Marshall Islanders, only nodules larger than 1 cm were counted.

In a study of Chernobyl cleanup workers conducted 1-8 years after exposure, 2.8 per 1,000 patients screened by ultrasound had cancer (Ivanov et al., 1997). Ultrasound screening for nodules 0.5 cm in diameter in Nagasaki survivors revealed cancer rates of 0.10-0.05 in the exposed verses the unexposed group, respectively (Nagataki et al., 1994).

Table F.3b lists the proportion of nodules found to be cancerous in exposed and unexposed populations. The proportion of cancers is higher for exposed populations and for larger nodules. In several studies, a nodule found by screening in exposed subjects was more likely to be malignant than was one found in an unexposed subject. In contrast to the probabilities seen in unexposed populations (0.01-0.05), nodules found in people exposed to radiation in the study of areas near the Nevada Test Site had a 0.27 probability of being malignant; in Nagasaki, this probability was 0.11 in the exposed group and 0.02 in the unexposed group. Absolute probabilities vary from study to study, but within studies a marked difference between exposed and unexposed individuals is observed consistently.

Size of Neoplasms Found Through Screening Exposed Populations

Because screening is intended to find cancers in an earlier stage than would be found in the course of usual care, the size of cancers found in a screening program is important. Early detection of thyroid cancers will improve survival only if the cancers found have a better prognosis than would have occurred if they were detected in usual care. No population-based study of individuals exposed to fallout reported this information. Only one other study, a recall program of medically irradiated people, reported sizes of thyroid cancers found (Favus et al., 1976). Of 60 cancers found by palpation and scintigraphy, 21 were under 0.5 cm, 28 were 0.6-1.5 cm, and 11 were larger than 1.5 cm. In a follow-up study of a subgroup—patients diagnosed before the start of the screening program—those with more clinically apparent symptoms had a higher risk of recurrence than did those diagnosed earlier, in the course of usual care (Schneider et al., 1986).

Stage and Histology of Cancers in Exposed Populations

No cases of undifferentiated thyroid cancer were reported in five population-based studies of exposed subjects (Howard et al., 1997; Inskip et al., 1997; Ivanov et al., 1997; Kerber et al., 1993; Takahashi et al., 1997). In a study of U.S. residents exposed to fallout radiation from the Nevada Test Site (Kerber et al., 1993), all cancers found were papillary. A study of Marshall Islanders (Takahashi et al., 1997) found 5 papillary cancers and 1 follicular cancer in exposed people. In 47 thyroid cancers detected in 167,862 Chernobyl cleanup workers (Ivanov et al., 1997), 42.8% were papillary, 33.3% were follicular, and 14.3% were other types of carcinoma. This last category is not explained further. Two cancers found in a smaller sample of Chernobyl workers (Inskip et al., 1997) were histologically confirmed to be papillary. None of these studies report information on other clinical features of the cancers found, such as lymph node involvement or distant metastasis.

Surveillance studies of medically irradiated subjects provide some information on histology of thyroid cancers found through screening. In one large series (Favus et al., 1976), all 60 thyroid cancers found in a surveillance program of patients exposed to radiation for benign conditions in childhood were papillary, follicular, or a combination of both (which would now be classified as papillary).

In 45 cases, one lobe was involved, and in 15, both lobes were. There was thyroid capsule invasion in 12 cases, blood vessel invasion in 5, and lymph node involvement in 17. No distant metastases were found. In two other surveillance studies that provide histologic information (Crom et al, 1997; Royce et al., 1979), all cancers found were papillary and follicular.

Probability of Undergoing Surgery

Unexposed Populations

Two studies of healthy, unexposed individuals provide information on surgery performed as a result of screening by palpation in unexposed populations. These results suggest that, of 1,000 unexposed people screened by palpation, between 1 and 8 will have surgery. In the United States, the Framingham study reports 45 surgical operations in 5,108 adults screened by palpation (Vander et al., 1968). None of the individuals proved to have a carcinoma. In another study of healthy Japanese workers, 1 person had surgery; 451 individuals were screened (Miki et al., 1993). This study was unusual in that the decision to proceed to FNA was made based on ultrasound diagnosis of features suggestive of cancer; as a result, very few FNAs and surgeries were performed relative to other studies.

Exposed Population

Studies of Marshall Islanders exposed to fallout radiation show much higher rates of surgery among the screened population. Hamilton and co-workers (1987) report 53 cases of surgery in 1,903 exposed Marshall Islanders screened by palpation. A decade later, in a subset of the same population, Howard and co-workers (1997) report 55 operations in 253 very highly exposed Marshall Islanders. Ten of 1,984 Chernobyl cleanup workers had surgery as a result of screening by palpation (Ivanov et al., 1997), but this study was done only a few years after the accident; continued monitoring could result in higher rates of surgery.

Recall studies of palpation screening in people who received head and neck irradiation in childhood show the highest rates of surgery among screened populations. For example, Shimaoka and co-workers (1982) report 85 operations as a result of palpation screening of 1,500 patients; Royce and co-workers (1979) report 6 in 214; Ron (1996) reports 8 in 443; Favus et al. (1976) report 195 operations in a study of 1,056 recalled patients. These surgery rates are about 10 times higher than those found as a result of screening of unexposed people.

Few data are available to estimate the false-negative rate for surgery in screening exposed populations. The rate can be estimated from information on the FNA specificity and from the ratio of surgery to cancers found in studies that report this information. In individuals exposed to fallout, approximately 1 in 3 patients undergoing surgery had cancer (Howard et al., 1997; Inskip et al., 1997). In patients who underwent external radiation for benign medical conditions, the proportion was 0.20 (Ron et al., 1984; Royce et al., 1979; Shimaoka et al., 1982).

Effectiveness of Screening

Screening in a population is justified if there is evidence that early detection and treatment before clinical symptoms are apparent will reduce mortality and morbidity or improve quality of life. Other issues to be considered include consequences of false-negative and false-positive test results, acceptability of the test to the screened population, and the risks of treatment.

There have been no randomized, controlled trials that address the question of whether screening for thyroid cancer improves patient outcomes. Because thyroid cancer is rare and has a long latency period, such trials are unlikely to be done in the future. Case-control studies can be used to judge the effectiveness of screening, but we could find no studies that evaluated screening programs using that model.

As shown in Figure F.1, stage is an important intermediate measure used in many studies as a substitute for direct evidence about mortality and morbidity. In theory screening is more likely to find a cancer that it is smaller, that has not yet metastasized, or that has metastasized less widely, than would be likely to be found without screening. In the absence of controlled studies, the only way to estimate the extent of a benefit would be to know how much earlier, compared to usual care, screening will detect cancers. In this context "earlier" means not only at an earlier time, but also at an earlier stage as judged by size, histology, or extent of spread. The studies of screening we reviewed contained no data on which to base such estimates.

Effect of Early Detection on Mortality

Three observational studies have attempted to address prognosis with reference to mortality. These papers show that given a particular size, cell type, and extent of spread—and controlling for age—the prognosis of cancers associated with exposure to radiation appears to be similar to that of cancers that occur in unexposed individuals (Schneider et al., 1986), that in an observational study of an unexposed population, the average size of cancers found by screening was smaller than that of a comparison, unscreened group (Ishida et al., 1988), and that in a statistically controlled study, delay in treatment was associated with lower survival-rate in an unexposed cohort (Mazzaferri and Jhiang, 1994).

In the observational study, Ishida and co-workers (1988) compared sizes of thyroid cancers found through screening with those found in the course of usual care. In the screened group, made up of healthy Japanese women without exposure to radiation, 58% of palpated tumors were 2 cm or smaller. The comparison group consisted of outpatients at a thyroid cancer clinic. In the clinic patients there were fewer cancers under 2 cm, and more tumors larger than 4 cm. In the screened group, all but 1 of 216 thyroid cancers were differentiated, and 85.5% were papillary. In the comparison group of 229 patients with thyroid cancer, 70.3% of the cancers were papillary, and there were 9 (3.9%) undifferentiated carcinomas. After 1.5-8 years of follow-up, 96.7% of screened patients were disease-free, compared with 85.6% in the comparison group. None of the screened patients died of thyroid cancer, but 12 patients (5.4%) in the clinic group did.

Flaws in the selection of the comparison group limit the usefulness of the article (Ishida et al., 1988). It is not clear whether the clinic patients had already been followed for many years. It appears that the investigators reported the sizes of the cancers in the clinic patients at the time of the study, rather than at the time of diagnosis, which could have been years earlier. The mean age of the clinic patients is given, but it is not clear whether the age distribution was similar in the screened group. Finally, the investigators apparently did not study whether cancers or cancer deaths occurred in screened individuals who were not diagnosed to have cancer on the initial and second screenings. Undifferentiated cancers that developed later in screened patients could have been excluded from the "screened" groups and, in fact, might even have been included in the "clinic" group. These flaws make comparison of tumor sizes and follow-up mortality meaningless.

In the statistically controlled study of a cohort of patients under the care of U.S. Air Force physicians, Mazzaferri and Jhiang (1994) record the time from the first manifestation of cancer to initial therapy. Patients who died of cancer had a median delay of 18 months, compared with 4 months for those who survived 30 years. A simple linear regression of the logarithm of delay in treatment on cancer death had a correlation coefficient of 0.49. The results suggest that a delay of 1 year increased the probability of dying of cancer by about 5 percentage points (from nearly 0 to 5%); a delay of 10 years instead of 1 year increased the probability of dying of cancer from 5% to approximately 12%.

What is the relevance of these findings to screening? First, the average tumor size was 2.5 cm, and the prognosis was inversely related to the size of the nodule at the time of diagnosis. Although the study did not directly estimate the rate of growth of untreated cancerous nodules, it stands to reason that earlier detection of these very large tumors could improve the outcome.

Second, the study documents that, in practice, treatment often was delayed for many years. The reasons are not made clear, and may have occurred after a clinical diagnosis was made. Long delays, followed by a poor outcome, could have been the result of the presence of malignant nodules that were inadvertently managed conservatively for many years, perhaps because an accurate diagnosis was not made initially. In any case, this finding argues (as the authors conclude) for more aggressive initial management of large nodules, once they are diagnosed, but it does not necessarily mean that diagnosing these nodules earlier than was done with usual care would substantially improve outcomes.

Third, the benefit of early detection was apparent only in patients with large nodules, and it should not be overgeneralized to patients with occult nodules or even smaller palpable nodules. If true, this finding suggests that screening be focused on finding large (>1.5 cm), potentially palpable, but clinically unsuspected nodules.

Effect of Screening on Morbidity

Stage at the time of detection, recurrence of cancer, and complications of surgery can affect morbidity (Figure F.1). There are no data from studies of screening about the effect of early detection on recurrence rates or on the frequency of surgical complications in the screened population.

Recurrence of Cancer

We found no studies that directly compare rates of recurrence in screened versus unscreened populations. Most screening studies involve one-time screening and do not follow patients over time for evidence of recurrence or other outcomes. One follow-up study of surveillance for thyroid cancer in patients exposed to external beam radiation for benign conditions of the head, neck, and thorax (Schneider et al., 1986) reports a lower recurrence rate in patients diagnosed after a screening program was begun compared with those diagnosed through usual care—reflecting factors such as smaller tumor size in cancers discovered through screening.

Factors associated with risk of recurrence, such as age at diagnosis (Krausz et al., 1993; Mazzaferri and Jhiang, 1994; Schindler et al., 1991; Schneider et al., 1986; Simpson, 1987; Tubiana et al., 1985), tumor size (Mazzaferri and Jhiang, 1994; Noguchi, 1995; Simpson, 1987; Schindler et al., 1991; Schneider et al., 1986), male sex (Noguchi et al., 1995; Schindler et al., 1991; Tubiana et al., 1985), and local invasion or metastases (Mazzaferri and Jhiang, 1994; Simpson et al., 1987; Schneider et al., 1986; Schindler et al., 1991; Simpson et al., 1987; Tubiana et al., 1985), are generally the same as those associated with increased mortality. Follow-up studies of thyroid cancer patients report recurrence rates of about 10-30%. About two-thirds of recurrences come within 10 years of initial diagnosis, but recurrence can occur throughout a patient's lifetime.

A thyroid cancer recurrence does not necessarily mean increased mortality (Krausz et al., 1993). Cure rates remain high even for recurrences, except for patients who have metastases at the time of recurrence (Mazzaferri and Jhiang, 1994). Few studies relate factors at initial diagnosis to mortality at recurrence, but those that do find survival of a recurrence affected most strongly by age. A Japanese study (Noguchi et al., 1995) with follow-up periods of 10-30 years examined risk factors in patients who died after recurrence with patients who survived a recurrence. Only age at initial diagnosis correlated with survival of a recurrence. Similarly, an Israeli study (Krausz et al., 1993) found only that age at initial diagnosis was a significant risk factor for mortality after recurrence. Another study of Japanese patients (Asakawa et al., 1997) found increased mortality of recurrent thyroid cancers to be associated with age at diagnosis and shorter period between initial diagnosis and recurrence. Tumor size and spread at initial diagnosis do not appear to be associated with mortality from recurrence.

Even though recurrent thyroid cancers are usually highly curable, Mazzaferri (1987) makes the point that a recurrence, even when cured, is never an insignificant event for the patient, in terms of psychological costs and quality of life.

Harms of Screening

Uncertainty about effectiveness is of particular concern when significant harms are associated with screening. Almost no information about quality of life or morbidity related to the diagnosis and treatment of thyroid cancer has been published. In general, false-positive test results can cause anxiety. Hypothyroidism, caused by thyroid ablation and treatment for hypothyroidism, unexpected complications of surgery or radioiodine therapy, and even preparation for tests used to find recurrences (Dow et al., 1997a,b), can cause morbidity and affect the quality of life of patients with thyroid cancer, but the consequences of most of these events and conditions have not been systematically studied.

The long-term complications of thyroid surgery are hypocalcemia, a serious, difficult-to-treat condition due attendant to hypoparathyroidism; and hoarseness, due to injury to the recurrent laryngeal nerve. The frequency of these complications depends on the extent of the procedure and the experience of the surgeon. Complication rates vary among centers (Fraker et al., 1997). For the most extensive procedure, total thyroidectomy, the risk of hypocalcemia range from 7% to 32% in four surgical series with long-term follow-up (Fraker et al., 1997). In the same study the risk of recurrent laryngeal nerve injury ranged from 0% to 7%. The study reported the experience of highly skilled surgeons at major centers. Major surgery of the thyroid is generally considered difficult, and results of experienced surgeons are thought to be much better than the "average" results of surgeons who perform thyroid surgery only occasionally. Less extensive surgery, such as "near total" and subtotal thyroidectomy, causes fewer complications, but may be associated with higher rates of tumor recurrence and mortality.

Summary

Table F.5 summarizes the findings of our review. The table shows estimates of relevant probabilities and of the number of significant events per 10,000 individuals screened. Many of the estimates, such as those for test performance and

TABLE F.5 Summary of Evidence Regarding the Benefits of Screening for Thyroid Cancer

	Probability	Palpation	Ultrasound	References
Probability of having a nodule				Table F.3a-b
Any size nodule	0.35			
0.5-1.0 cm	0.12			
1.0-1.5 cm	0.05			
>1.5 cm	0.02			
Probability that a nodule is cancer				Table F.3a-b
nodules smaller than 1 cm	0.03			
nodules larger than 1 cm	0.10			
Cut-off for a significant nodule		1.0 cm	0.5 cm	Figure F.2
Probability of cancer given cut-off		0.0007	0.011	
Sensitivity for nodules				Tables F.1, 4
0.5-1.0 cm		0.00	1	
1.0-1.5 cm		0.55	1	
>1.5 cm		0.65	1	
Specificity for nodules				Tables F.1, 4
0.5-1.0 cm			0.98	
1.0-1.5 cm		0.98	0.99	
>1.5 cm		0.98	1.00	
Fine needle aspiration (FNA)				Table F.2a-b
Sensitivity for cancer	0.8			
False positive rate	0.35			
Lobectomy or subtotal thyroidectomy				Fraker et al.
Complication (recurrent laryngeal nerve injury)	0.0005			
Complication (hypocalcemia)	0.0005			
Total thyroidectomy				Fraker et al.
Complication (recurrent laryngeal nerve injury)	0.04			
Complication (hypocalcemia)	0.1			
Morbidity and mortality prevented by screening		—	—	No data
Summary				
Events per 10,000 screened				
Diagnosed to have a nodule (true positives)		405	1900	
Diagnosed to have a nodule (false positives)		386	271	
Undergo at least one FNA		791	2171	
Undergo FNA but do not have cancer		751	2076	
Undergo lobectomy		102	696	
Cancers diagnosed		32	85	
Cancers missed		38	35	
Surgical complications		5	13	
Death prevented in 5 years		No data	No data	

surgical complication rates, are derived from studies performed in specialized settings that are not likely to be representative of the results of a program of screening in the general population.

Data were insufficient to estimate the effect of screening on mortality or morbidity from thyroid cancer and from recurrences of thyroid cancer. The information needed to make such a determination involves detailing survival rates by risk category among patients found to have cancer during the course of usual care and estimating the effect of early detection on the distribution cancers among people according to category.

As shown in Table F.6a, the SEER study provides estimates of 5-year survival by sex and by the extent of spread (localized, regional, or distant) found among cancers diagnosed in usual care. It also provides estimates of the proportion of cancers found in each risk category. These data can be used to estimate overall survival for males and females with thyroid cancer found in the general population. Although the SEER data provide incomplete information about risk (there are no data by histologic type, for example), these are the only population-based data available about the distribution through risk categories of cancers found during usual care.

To determine whether early detection improves outcome, it would be necessary to estimate how often cancers are detected earlier, and how much earlier they are detected. Specifically, it would be necessary to know the average delay in diagnosis through usual care compared with that from screening and the rates of growth or spread of the cancers between the time they would be found by screening and the time they would be found in the usual course of care.

Table F.6b shows data elements—but not actual data—in one possible causal chain linking earlier diagnosis to improved survival. The improvement in survival could be estimated if we knew the distribution of tumor sizes in the usual-care population and in a screened population; the average duration of delay in diagnosis in usual care, relative to screening; the rate at which tumors in one size category progress to a larger size category; and the relationship between the size of the cancer and the likelihood that the cancer is localized. Because data about these variables are lacking—studies of screening have not demonstrated to what degree the cancers found by screening have a better prognosis—we could not make a reliable estimate of the benefits of screening.

How large is such a benefit likely to be? Table F.6c depicts a hypothetical example showing the potential relationship between earlier detection and 5-year survival for adult females. As shown in Table 6a, in the general population 60% of thyroid cancers were localized at the time of detection; 31% had spread to regional nodes. If screening increased the proportion of cancers that had no regional or distant metastases from 60% to 70%, the overall 5-year survival for thyroid cancer would increase from 0.947 to 0.952. Put differently, 184 people with thyroid cancer would need to be treated to prevent 1 death in 5 years. As seen in Table F.5, 32 cancers would be found per 10,000 people screened. There-

TABLE F.6 Hypothetical Relationship Between Efficacy of Early Detection and Five-Year Survival from Thyroid Cancer

a. SEER data

Risk categories	Five-year survival			Proportion of cancers in each risk category	Overall Survival	
	Total	Males	Females		Males	Females
localized	0.997	0.994	0.998	0.60	0.596	0.599
regional	0.937	0.913	0.947	0.31	0.283	0.294
distant	0.448	0.403	0.475	0.05	0.020	0.024
unstaged	0.766	0.773	0.761	0.04	0.031	0.030
5 year and overall survival				**0.931**	**0.947**	

b. Consequences of screening (categories of information needed to estimate effect of screening on dectection of disease by stage)

	Usual Care	Screening	
size categories			
(1) 0.5 -1.0 cm			%
(2) 1.0- 1.5 cm			%
(3) 1.5 cm or larger			%
delay in diagnosis (average)			years
rates of transitions between size categories			
(1) to (2)			% per year
(2) to (3)			% per year
relation of size to spread of cancer			
(1) 0.5-1.0 cm			% localized
(2) 1.0-1.5 cm			% localized
(3) 1.5 cm or larger			% localized

c. Hypothetical improvement in survival in relation to risk categories if screening increased proportion of cancers with no metasteses from 60% to 70% compared to usual care (adult females only)

	Usual Care	Screening
Risk categories		
localized	0.6	0.7
regional	0.31	0.21
distant	0.05	0.05
unstaged	0.04	0.04
Contribution to 5 year survival	0.947	0.952

Number needed to treat to prevent 1 death in 5 years	184
Number needed to screen to prevent 1 death in 5 years	57,445

fore, 1 death would be prevented for every 57,445 individuals screened. Screening 57,000 individuals could also result in 26 surgical complications (Table F.5). The implication is that, in the future, studies of screening should attempt to clearly document the stages at which cancers are detected, and an appropriate comparison group should be identified to determine whether screening is beneficial.

References

Antonelli A, Silvano G, et al. 1995. Risk of thyroid nodules in subjects occupationally exposed to radiation: a cross sectional study [see comments]. Occupational & Environmental Medicine, 52(8):500-504.

Asakawa H, Kobayashi T, et al. 1997. Prognostic factors in patients with recurrent differentiated thyroid carcinoma. Journal of Surgical Oncology 64(3):202-206.

Ashcraft MW, Van Herle AJ. 1981. Management of thyroid nodules. II: Scanning techniques, thyroid suppressive therapy, and fine needle aspiration. Head & Neck Surgery 3(4):297-322.

Bouvet M, Feldman IJ, et al. 1992. Surgical management of the thyroid nodule: patient selection based on the results of fine-needle aspiration cytology. Laryngoscope 102(12 Pt 1):1353-6.

Brander A, Viikinkoski P, Nickels J, Kivisaari L. 1989. Thyroid gland: US screening in middle-aged women with no previous thyroid disease. Radiology, 173(2):507-510.

Brander A, Viikinkoski P, Nickels J, Kivisaari L. 1991. Thyroid gland: US screening in a random adult population. Radiology, 181(3):683-687.

Brander A, Viikinkoski P, Tuuhea J, Voutilainen L, Kivisaari L. 1992. Clinical versus ultrasound examination of the thyroid gland in common clinical practice. Journal of Clinical Ultrasound, 20(1):37-42.

Bruneton J, Balu-Maestro C, Marcy P, Melia P, Mourou M. 1994. Very high frequency (13 Mhz) ultrasonographic examination of the normal neck: detection of normal lymph nodes and thyroid nodules. Journal of Ultrasound Medicine 13:87-90.

Carroll, BA. 1982. Asymptomatic thyroid nodules: incidental sonographic detection. American Journal of Roentgenology 138(3):499-501.

Caruso, D, Mazzaferri E L. 1991. Fine needle aspiration biopsy in the management of thyroid nodules. The Endocrinologist 1:194-202.

Crom D B, Kaste SC, et al. 1997. Ultrasonography for thyroid screening after head and neck irradiation in childhood cancer survivors. Medical & Pediatric Oncology 28(1):15-21.

Cusick EL, MacIntosh CA, et al. 1990. Management of isolated thyroid swellings: a prospective six year study of fine needle aspiration cytology in diagnosis. British Medical Journal 301(6747): 318-21.

Dow KH, Ferrell BR, AnelloC. 1997a. Balancing demands of cancer surveillance among survivors of thyroid cancer. Cancer Pract., 5(5):289-95.

Dow KH, Ferrell BR, Anello C. 1997b. Quality-of-life changes in patients with thyroid cancer after withdrawal of thyroid hormone therapy. Thyroid., 7(4):613-69.

Eddy D 1991. "How to think about screening," in Eddy, D. ed, Common Screening Tests. Philadelphia: American College of Physicians, pp. 1-10.

Ezzat S, Sarti DA, Cain DR, Braunstein GD. 1994. Thyroid incidentalomas. Prevalence by palpation and ultrasonography. Archives of Internal Medicine, 154(16):1838-1840.

Favus MJ, Schneider AB, et al. 1976. Thyroid cancer occurring as a late consequence of head-and-neck irradiation. Evaluation of 1056 patients. N Engl J Med, 294(19):1019-1025.

Fraker DL, Skarulis M, Livolsi V. 1997. Thyroid Tumors. In DeVita, Jr., VT, Hellman, S, Rosenberg, SA., eds. Cancer: principles and practice of oncology. Philadelphia: Lippincott-Raven, pp. 1629-1652.

Gharib H. 1994. Fine-needle aspiration biopsy of thyroid nodules: advantages, limitations, and effect. Mayo Clinic Proceedings, 69(1):44-49.

Gharib H, Goellner JR. 1993. Fine-needle aspiration biopsy of the thyroid: an appraisal. Annals of Internal Medicine, 118(4):282-289.

Giuffrida D, Gharib H. 1995. Controversies in the management of cold, hot, and occult thyroid nodules. American Journal of Medicine, 99(6):642-650.

Haas S, Trujillo A, et al. 1993. Fine needle aspiration of thyroid nodules in a rural setting. American Journal of Medicine 94(4):357-61.

Hall TL, Layfield LJ, Philippe A, Rosenthal DL. 1989. Sources of diagnostic error in fine needle aspiration of the thyroid. Cancer, 63(4):718-725.

Hamburger JI, Hamburger SW. 1985. Declining role of frozen section in surgical planning for thyroid nodules. Surgery 98(2):307-12.

Hamilton TE, van Belle G, LoGerfo JP. 1987. Thyroid neoplasia in Marshall Islanders exposed to nuclear fallout. Jama, 258(5):629-635.

Holleman F, Hoekstra JB, et al. 1995. Evaluation of fine needle aspiration (FNA) cytology in the diagnosis of thyroid nodules. Cytopathology 6(3):168-75.

Horlocker TT, Hay JE, et al. 1986. Prevalence of incidental nodular thyroid disease detected during high-resolution parathyroid ultrasonography. In Medeiros-Neto J, Gaitan E (eds.), Frontiers in Thyroidology. Vol. II. Plenum Publishing Corp. pp. 1309-1312.

Howard JE, Vaswani A, Heotis P. 1997. Thyroid disease among the Rongelap and Utirik population—an update. Health Physics, 73(1):190-198.

Hsiao YL, Chang TC. 1994. Ultrasound evaluation of thyroid abnormalities and volume in Chinese adults without palpable thyroid glands. Journal of the Formosan Medical Association, 93(2):140-144.

Inskip PD, Hartshorne MF, et al. 1997. Thyroid nodularity and cancer among Chernobyl cleanup workers from Estonia. Radiation Research 147(2):225-35.

Ishida T, Izuo M, Ogawa T, Kurebayashi J, Satoh K. 1988. Evaluation of mass screening for thyroid cancer. Japanese Journal of Clinical Oncology, 18(4):289-295.

Ivanov VK, Tsyb AF, et al. 1997. Leukaemia and thyroid cancer in emergency workers of the Chernobyl accident: estimation of radiation risks (1986-1995). Radiation & Environmental Biophysics 36(1): 9-16.

Jones AJ, Aitman TJ, et al. 1990. Comparison of fine needle aspiration cytology, radioisotopic and ultrasound scanning in the management of thyroid nodules. Postgrad Med J, 66(781):914-917.

Kerber RA, Till JE, Simon SL, et al. 1993. A cohort study of thyroid disease in relation to fallout from nuclear weapons testing. JAMA 270(17):2076-2082.

Khafagi F, Wright G, et al. 1988. Screening for thyroid malignancy: the role of fine-needle biopsy. Medical Journal of Australia 149(6):302-3, 306-7.

Krausz Y, Uziely B, et al. 1993. Recurrence-associated mortality in patients with differentiated thyroid carcinoma. Journal of Surgical Oncology 52(3):154-8.

Kuma K, Matsuzuka F, Yokozawa T, Miyauchi A, Sugawara M. 1994. Fate of untreated benign thyroid nodules: results of long-term follow-up. World J Surg. 18:495-8. Cited in Tan and Gharib, 1997.

Landis SH, Murray T, Bolden S, Wingo PA. 1998. Cancer statistics, 1998. CA Cancer J Clin, 48(1):6-29.

Lin JD, Huang BY, et al. 1997. Thyroid ultrasonography with fine-needle aspiration cytology for the diagnosis of thyroid cancer. Journal of Clinical Ultrasound 25(3):111-8.

Mazzaferri EL. 1987. Papillary thyroid carcinoma factors influencing prognosis and current therapy [published erratum appears in Semin Oncol 1988 Jun; 15(3):X]. Seminars in Oncology 14(3): 315-32.

Mazzaferri EL, Jhiang SM. 1994. Long-term impact of initial surgical and medical therapy on papillary and follicular thyroid cancer. [see comments] [published erratum appears in Am J Med 1995 Feb. 98(2):215]. American Journal of Medicine 97(5):418-28.

McTiernan AM, Weiss NS, Daling JR. 1984. Incidence of thyroid cancer in women in relation to previous exposure to radiation therapy and history of thyroid disease. J Natl Cancer Inst, 73(3):575-581.

Merchant WJ, Thomas SM, et al. 1995. The role of thyroid fine needle aspiration (FNA) cytology in a District General Hospital setting. Cytopathology 6(6):409-18.

Mettler FA, Williamson MR, et al. 1992. Thyroid nodules in the population living around Chernobyl. JAMA 268(5):616-9.

Miki H, Oshimo K, et al. 1993. Incidence of ultrasonographically-detected thyroid nodules in healthy adults. Tokushima Journal of Experimental Medicine, 40(1-2):43-46.

Moosa M, Mazzaferri EL. 1997. Occult thyroid carcinoma. The Cancer Journal, 10, 180-188.

Morayati SJ, Freitas JE. 1991. Guiding thyroid nodule management by fine-needle aspiration. Family Practice Research Journal 11(4):379-86.

Nagataki S. Shibata Y, et al. 1994. Thyroid diseases among atomic bomb survivors in Nagasaki [published erratum appears in JAMA 1995 Jan 25;273(4):288]. JAMA 272(5): 64-70.

Ng EH, Tan SK, et al. 1990. Impact of fine needle aspiration cytology on the management of solitary thyroid nodules. Australian & New Zealand Journal of Surgery 60(6):463-6.

Noguchi M, Yagi H, et al. 1995. Recurrence and mortality in patients with differentiated thyroid carcinoma. International Surgery 80(2):162-6.

Ongphiphadhanakul B, Rajatanavin R, et al. 1992. Systematic inclusion of clinical and laboratory data improves diagnostic accuracy of fine-needle aspiration biopsy in solitary thyroid nodules. Acta Endocrinologica 126(3):233-7.

Piromalli D, Martelli G, et al. 1992. The role of fine needle aspiration in the diagnosis of thyroid nodules: analysis of 795 consecutive cases. Journal of Surgical Oncology 50(4):247-50.

Pottern LM, Kaplan MM, et al. 1990. Thyroid nodularity after childhood irradiation for lymphoid hyperplasia: a comparison of questionnaire and clinical findings. Journal of Clinical Epidemiology, 43(5):449-460.

Ries LAG, Kosary CL, et al. eds. 1997. SEER Cancer Statistics Review, 1973-1994, National Cancer Institute. NIH Pub. No. 97-2789. Bethesda, MD.

Ron E 1996. Thyroid Cancer. In Shottenfeld, D., Fraumeni, Jr., JF, eds. Cancer Epidemiology and Prevention. 2nd Edition. New York: Oxford University Press, pp. 1000-1021.

Ron E, Lubin E, Modan B. 1984. Screening for early detection of radiation-associated thyroid cancer: a pilot study. Israel Journal of Medical Sciences, 20(12):1164-1168.

Ron E, Kleinerman RA, et al. 1987. A population-based case-control study of thyroid cancer. J Natl Cancer Inst, 79(1):1-12.

Ron E, Modan B, Boice JDJ. 1988. Mortality after radiotherapy for ringworm of the scalp. Am J Epidemiol, 127(4):713-725.

Ron E, Modan B, et al. 1989. Thyroid neoplasia following low-dose radiation in childhood. Radiation Research, 120(3):516-531.

Ron E, Lubin JH, et al. 1995. Thyroid cancer after exposure to external radiation: a pooled analysis of seven studies. Radiation Research, 141(3):259-277.

Ron E, Saftlas AF. 1996. Head and neck radiation carcinogenesis: epidemiologic evidence. Otolaryngology—Head & Neck Surgery, 115(5):403-408.

Royce PC, MacKay BR, et al. 1979. Value of postirradiation screening for thyroid nodules. A controlled study of recalled patients. JAMA 242(24):2675-8.

Schindler AM, van Melle G, et al. 1991. Prognostic factors in papillary carcinoma of the thyroid. Cancer 68(2):234-30.

Schlumberger MJ. 1998. Papillary and follicular thyroid carcinoma. NEJM 338:297-306.

Schneider AB, Recant W, et al. 1986. Radiation-induced thyroid carcinoma. Clinical course and results of therapy in 296 patients. Annals of Internal Medicine 105(3):405-412.

Schneider AB, Ron E, Lubin J, Stovall M, Gierlowski TC. 1993. Dose-response relationships for radiation-induced thyroid cancer and thyroid nodules: evidence for the prolonged effects of radiation on the thyroid. Journal of Clinical Endocrinology & Metabolism 77(2):362-369.

Schneider AB, Bekerman C, Leland J. 1997. Thyroid nodules in the follow-up of irradiated individuals: comparison of thyroid ultrasound with scanning and palpation. J Clin Endoc Metab 82:4020-4027.

Shimaoka K, Bakri K, et al. 1982. Thyroid screening program; follow-up evaluation. New York State Journal of Medicine, 82(8):1184-1187.

Shore, R. E., Hildreth, N., et al. (1993). Thyroid cancer among persons given X-ray treatment in infancy for an enlarged thymus gland. American Journal of Epidemiology, 137(10):1068-1080.

Shore RE. 1992. Issues and epidemiologycal evidence regarding radiation-induced thyroid cancer. Radiation Research 1312(1):98-111.

Shore RE, Hildreth N, et al. 1993. Thyroid cancer among persons given X-ray treatment in infancy for enlarged thymus gland. American Journal of Epidemiology, 137(10):1068-1080.

Simpson WJ, McKinney SE, et al. 1987. Papillary and follicular thyroid cancer. Prognostic factors in 1,578 patients. American Journal of Medicine 83(3):479-88.

Spiliotis J, Scopa CD, et al. 1991. Diagnosis of thyroid cancer in southwestern Greece. Bulletin du Cancer 78(10):953-9.

Takahashi T, Trott KR, et al. 1997. An investigation into the prevalence of thyroid disease on Kwajalein Atoll, Marshall Islands. Health Physics 73(1):199-213.

Takashima S, Fukuda H, et al. 1994. Thyroid nodules: clinical effect of ultrasound-guided fine-needle aspiration biopsy. Journal of Clinical Ultrasound 22(9):535-42.

Tan GH, Gharib H, Reading CC. 1995. Solitary thyroid nodule. Comparison between palpation and ultrasonography. Archives of Internal Medicine, 155(22):2418-2423.

Tan GH, Gharib H. 1997. Thyroid incidentalomas: management approaches to nonpalpable nodules discovered incidentally on thyroid imaging. Annals of Internal Medicine, 126(3):226-231.

Tubiana M, Schlumberger M, et al. 1985. Long-term results and prognostic factors in patients with diffrentiated thyroid carcinoma. Cancer 55(4):794-804.

Tucker MA, Jones PH, et al. 1991. Therapeutic radiation at a young age is linked to secondary thyroid cancer. The Late Effects Study Group. Cancer Research, 51(11):2885-2888.

Tunbridge WM, Evered DC, et al. 1977. The spectrum of thyroid disease in a community: the Whickham survey. Clinical Endocrinology, 7(6):481-493.

USDHHS (U. S. Department of Health and Human Services, National Cancer Institute). 1997. Estimated exposures and thyroid doses received by the American people from iodine-131 in fallout following Nevada atmospheric nuclear bomb test. NIH Pub. 97-4264.

Vander JB, Gaston EA, Dawber TR. 1954. The significance of nontoxic thyroid nodules. Premiminary report. New England Journal of Medicine 251(24):970-973.

Vander JB, Gaston EA, Dawber TR. 1968. The significance of nontoxic thyroid nodules. Final report of a 15-year study of the incidence of thyroid malignancy. Annals of Internal Medicine, 69(3):537-540.

Watters DA, Ahuja AT, et al. 1992. Role of ultrasound in the management of thyroid nodules. American Journal of Surgery 164(6):654-7.

Wingo PA, Tong T, et al. 1995. Cancer statistics, 1995 [published erratum appears in CA Cancer J Clin 1995 Mar-Apr; 45(2):127-8]. CA: a Cancer Journal for Clinicians 45(1):8-30.

Woestyn J, Afschrift M, Schelstraete K, Vermeulen A. 1985. Demonstration of nodules in the normal thyroid by echography. British Journal of Radiology 58(696):1179-1182.

Committee Biographies

COMMITTEE ON THYROID SCREENING RELATED TO I-131 EXPOSURE

ROBERT LAWRENCE, M.D., *Co-Chair,* is the Associate Dean for Professional Education and Programs and Professor of Health Policy at Johns Hopkins University School of Hygiene and Public Health. Dr. Lawrence is a member of the Institute of Medicine and currently serves as the chair of the Committee on Priorities for Vaccine Development. He recently chaired the Committee on Health Services in the U.S.-Associated Pacific Basin. Prior committee memberships have included: IOM Committee on Health and Human Rights (chair), Committee on Human Rights of the NAS, NAE, and IOM, Subcommittee to Evaluate NASA Medical Surveillance Data Sheets, and the Board on Health Promotion and Disease Prevention (chair). His expertise and research interests include community and social medicine, human rights, health promotion and disease prevention, evidence-based decision rules for prevention policy, and international health.

CATHERINE BORBAS, PH.D., Executive Director of the Healthcare Education and Research Foundation in St. Paul, Minnesota is an expert in the area of managed care and practice guidelines. Prior committee membership includes the Committee on Methods for Setting Priorities for Guidelines for the Division of Health Care Services of the Institute of Medicine. Dr. Borbas has published in the areas of clinical guidelines methodology, assessment, and implementation. Dr. Borbas earned her Ph.D. in social work and masters in public health from the University of Minnesota.

J. WILLIAM CHARBONEAU, M.D., is Professor of Radiology at the Mayo Medical School in Rochester Minnesota. He is also a staff physician at the Mayo Clinic Department of Radiology. He is a member of the American Medical Association, American College of Radiology, American Roentgen Ray Society, Radiologic Society of North America, Society of Gastrointestinal Radiology, Society of Radiologists in Ultrasound, and Zumbro Valley Medical Society. He has published 100 articles in professional journals since 1976. He was the coeditor for two textbooks on ultrasonography and two course syllabi on sonography. He was also the assistant editor for the *Mayo Clinic Family Health Book* published in 1990. Dr. Charboneau received his M.D. from the University of Wisconsin where he also completed graduate work in anatomy. He did his residency at the Mayo Graduate School of Medicine. He is licensed to practice medicine in Minnesota, Iowa and Wisconsin, and he is American board certified in radiology.

VIRGINIA A. LIVOLSI, M.D., is Professor of Pathology in the Department of Pathology and Laboratory Medicine at the University of Pennsylvania School of Medicine. She has received the Medal of Honor from Tokyo University in Tokyo, Japan. She is a member of the National Action Plan on Breast Cancer: Biological Resources Bank Working Group and served a member of the review committee of the Specialized Programs for Research Excellence in Breast Cancer (1992). Since 1990 she has written 15 articles and papers on various aspects of thyroid cancer for numerous publications including *Endocrine Pathology, American Journal of Clinical Pathology, Clinical Oncology, Modern Pathology, Human Pathology,* and *Cancer.* Dr. LiVolsi received her M.D. from Columbia University College of Physicians and Surgeons. She interned in pathology at Columbia-Presbyterian Medical Center in New York and she worked in a National Cancer Institute Traineeship in Surgical Pathology as well as served as the Chief Resident in Pathology at Columbia-Presbyterian Medical Center.

ERNEST L. MAZZAFERRI, M.D., M.A.C.P., Professor and Chairman of the Department of Internal Medicine at The Ohio State University in Columbus, OH is internationally known for his work in the treatment of thyroid cancer. He is the former chair of the American College of Physicians' Clinical Efficacy Assessment Subcommittee and has published articles on screening and treatment effectiveness.

STEPHEN G. PAUKER, M.D. Vice Chairman for Clinical Affairs, New England Medical Center, and Professor of Medicine, Tufts University is an expert on clinical decision making and evidence-based medicine. Dr. Pauker is a member of the Institute of Medicine. Previously, he has served on the Committee to Evaluate the Artificial Heart Program of the National Heart, Lung, and Blood Institute and the Workshops on the National Institutes of Health Consensus Development Process and the Use of Drugs in the Elderly, both within the IOM. His publications and research have addressed decisions about screening for cancer and other conditions. Dr. Pauker earned his medical degree at Harvard University in 1968 and trained in internal medicine and cardiology at Boston City and Massachusetts General Hospitals and the New England Medical Center, all in Boston.

HENRY ROYAL, M.D., is the Associate Director of the Division of Nuclear Medicine at the Mallinckrodt Institute of Radiology in St. Louis and is a Professor of Radiology at the Washington University School of Medicine. Dr. Royal earned his M.D. from St. Louis University. He is board-certified in internal medicine and in nuclear medicine. In 1990, Dr. Royal was the Co-Team Leader of the Health Effects Portion of the IAEA's International Chernobyl Project. In 1994-1995, he was a member of the Presidential Committee on Human Radiation Experiments. Dr. Royal is currently a Council Member of the National Council of Radiation Protection and Measurements and is a member of the Veterans Advisory Committee on Environmental Hazards.

SAMUEL A. WELLS, JR., M.D., is the Bixby Professor of Surgery and Chairman of the Department of Surgery at the Washington University School of Medicine, St. Louis, Missouri. Dr. Wells is a member of the Institute of Medicine, the American Association for Cancer Research, the American College of Surgeons (Fellow), the American Society of Clinical Oncology, the American Society for Clinical Investigation, the American Surgical Association, the Association of American Physicians, the Endocrine Society, the Society for Surgical Oncology, and several other professional societies. He has served on national and international committees for the General Motors Cancer Research Foundation, the John A. Hartford Foundation, the Lasker Award Jury, the National Institutes of Health, and the Societe Internationale de Chirurgie.

STEVEN H. WOOLF, M.D., Professor of Family Medicine at the Department of Family Practice at the Medical College of Virginia, is proposed as an expert on the development of clinical practice guidelines and on evidence-based medicine. Dr. Woolf has served on the Committee on Children, Health Insurance, and Access to Care created by the Board on Children, Youth and Families of the Institute of Medicine. He practices Family Medicine at the Fairfax Family Practice Center, Fairfax, Virginia. As Scientific Advisor to the U.S. Preventive Services Task Force from 1987 to 1996, Dr. Woolf has special expertise in the evaluation of screening tests. Dr. Woolf earned his medical degree at Emory University School of Medicine in 1984, interned at George Washington University Hospital in Washington, DC, did a residency in preventive medicine at Johns Hopkins University Hospital in Baltimore, MD and a residency in family practice at the Medical College of Virginia in Fairfax, VA.

COMMITTEE ON EXPOSURE OF THE AMERICAN PEOPLE TO
I-131 FROM THE NEVADA ATOMIC BOMB TESTS

WILLIAM J. SCHULL, PH.D., *Co-Chair*, is the Director of the Human Genetics Center at the University of Texas' School of Public Health. His specialty is human genetics and his primary research interest is radiation biology. In addition to a distinguished academic career, Dr. Schull previously served as the Head of the Department of Genetics with the Atomic Bomb Casualty Commission (ABCC) in Japan and later went on to become one of the Directors of the ABCC's successor organization, the Radiation Effects Research Foundation. Dr. Schull is the recipient of numerous awards, including the Order of the Sacred Treasure, Third Class. Dr. Schull is the member of several professional societies including the American Epidemiological Society, the American Society of Human Genetics, the Radiation Research Society, and the Society for the Study of Human Biology.

KEITH BAVERSTOCK, PH.D., a graduate of London University, has led the Radiation Protection Programme at the World Health Organization European Centre for Environment and Health in Rome, Italy since its foundation in 1991. This program was instrumental in bringing to world attention the increase in thyroid cancer in Belarus, now attributed to the Chernobyl accident. Prior to 1991 Dr. Baverstock was at the UK Medical Research Council Radiobiology Unit at Harwell, UK where he pursued a wide range of scientific research interests related to the public and occupational health aspects of exposure to ionizing radiation. He managed and analyzed the survey of UK radium luminizers, which has provided direct information on the effects of exposure to ionizing radiation at low dose rates. He held a visiting research appointment at the Institute of Chemical and Physical Sciences, RIKEN, Wako-shi, Japan, working on the effects of heavy ion radiation of DNA. During his tenure with the Medical Research Council Dr. Baverstock served as secretary to, and served upon a number of committees charged with advising the Council on the biological bases of the effects of ionizing radiation on man. Also during this tenure Dr. Baverstock served on the oversight committee for the Nation-wide Radiological Survey of the Marshall Islands and was Chairman of the Scientific Management Team of the Rongelap Resettlement Project.

STEPHEN BENJAMIN, PH.D., Colorado State University, Fort Collins, Colorado. Dr. Benjamin is Professor in the Department of Pathology, Radiological Health Sciences, and Environmental Health in the College of Veterinary Medicine and Biomedical Sciences at Colorado State University. He is also Co-director of the Center for Environmental Toxicology and Technology at Colorado State. He earned his D.V.M. and Ph.D. degrees at Cornell University and is board-certified in veterinary pathology. He was formerly the associate dean of the Graduate School and director of the Collaborative Radiological Health Laboratory at Colorado State University. He has served on the National Academy of Sciences Committee on Assessment of Center for Disease Control Radiation Studies and on the Subcommittee on Liver Risk of the National Council on Radiation Protection and Measurements. Dr. Benjamin's research expertise is in experimental carcinogenesis with an emphasis on radiation and environmental chemicals, including thyroid carcinogenesis.

PATRICIA BUFFLER, PH.D., is Dean and Professor of Epidemiology at the University of California, Berkeley School of Public Health. Her current research interests in epidemiology include studies of leukemia in children, health effects of environmental tobacco smoke, and health effects of non-ionizing radiation. She has served on numerous national and international advisory groups including advisory committees to the Department of Energy, the Department of Defense, the Department of Health and Human Services, the Environmental Agency, the Office of the President, the National Research Council and the World Health Organization. Since 1996 she has served as a Visiting Director for the US-Japan Radiation Effects Research Foundation. She has served as President for the Society of Epidemiologic Research, the American College of Epidemiology, and the International Society for Environmental Epidemiology and is currently an officer of the Medical Sciences Council of the American Association for the Advancement of Science. She was awarded the American College of Epidemiology Lilienfeld Award in 1996. She is a Fellow of both the American College of Epidemiology and the Association for the Advancement of Science and a member of the Institute of Medicine/National Academy of Sciences.

SHARON DUNWOODY, PH.D., University of Wisconsin-Madison, Madison, WI. Dr. Sharon Dunwoody is Evjue-Bascom Professor of Journalism and Mass Communication at the University of Wisconsin-Madison. She also serves as Chair of Academic Programs for the University's Institute for Environmental Studies and as Head of the Center for Environmental Communication and Education Studies (CECES). She earned her Ph.D. in mass communication at Indiana University before moving to UW-Madison in 1981. She studies the role of the mass media in public understanding of science and has authored numerous journal articles, book chapters, and books on the topic. She is a member of the Committee on the Public Understanding of Science and Technology of the American Association for the Advancement of Science and a member of the NRC's Commission on Life Sciences.

PETER GROER, PH.D., is Associate Professor in the Department of Nuclear Engineering at the University of Tennessee-Knoxville. Dr. Groer earned his Ph.D. in Theoretical Physics from the University of Vienna, Austria. He teaches undergraduate and graduate courses in radiation protection, radiation risk, and reliability analysis. His research interests include Bayesian methods for radiation detection, dosimetry and risk and reliability analysis. He has served on the editorial board of *Risk Analysis*. He was a member of the NRC Committee on the Health Effects of Radon and Other Internally Deposited Alpha Emitters (BEIR IV) and served on several scientific committees of the National Council on Radiation Protection and Measurements. He is presently a member of the EPA Science Advisory Board's Uncertainty in Radiogenic Risks Subcommittee. He is an avid tennis player and was on Austria's Olympic Basketball Team in 1960 and 1964.

ROBERT LAWRENCE, M.D. (see above).

CARL MANSFIELD, M.D., is Chairman of the Radiation Oncology Department at the University of Maryland Medical Systems. His research interest has been in the treatment of cancer with emphasis on breast cancer. Dr. Mansfield has done extensive research in radiation dosimetry and brachytherapy. From 1976 to 1983, Dr. Mansfield was Chairman of the Department of Radiation Oncology at the University of Kansas. From 1983 through 1995, Dr. Mansfield was Professor and Chairman of the Department of Radiation Oncology and Nuclear Medicine at Thomas Jefferson University Hospital. From 1995 to 1997, he was the Associate Director of the Radiation Research Program at the National Cancer Institute. Dr. Mansfield is a Fellow of the American College of Radiology, the American College of Nuclear Medicine, and the Philadelphia College of Physicians. Dr. Mansfield has served on committees for the National Cancer Institute and the National Research Council.

JAMES MARTIN, PH.D., is an Associate Professor of Radiological Health at the University of Michigan. His undergraduate degree is in physics and he received the MPH (1961) and Ph.D. (1965) degrees in radiological health from the University of Michigan. He is a certified Health Physicist (ABHP) and his research, teaching, and service is in radiation protection, radiation physics, radioactive waste management, radiological assessment, radioanalytical measurements, internal radiation dosimetry, and radiation protection standards. He is Project Director, International Low-level Radiation Waste Research and Education Institute and Director of the Michigan Radon Research and Training Center at the university. From 1979 to 1981 he managed Colorado's hazardous and solid waste program. From 1957 to 1978, he was with the USPHS and EPA where his duties involved environmental studies for reactors and nuclear fallout, technology assessments, and development of uranium fuel cycle standards and federal x-ray guides (issued by the President). He also directed EPA's program on radwaste standards. He served as a recent member of the Secretary of Energy's Advisory Committee on Nuclear Facility Safety for review of DOE facilities, and as a member of EPA's Science Advisory Board, Radiation Advisory Committee. He was the former Chair and Commissioner of the Michigan Toxic Substance Control Commission and was a member of the Governor's Task Force on High Level

Radioactive Waste, a panelist on EPA's workshop to revise national drinking water standards for radioactivity, and a member of the external review committee, Los Alamos National Laboratory Weapons Division. He is currently a member of the National Advisory Committee on Environmental Policy and Technology WIPP Subcommittee and a member of the NAS/NRC Committee on CDC Dose Reconstruction Studies for DOE facilities, and current Chair of the DOE Environmental Management Advisory Board (EMAB) Committee to develop guiding principles for remediation of old Manhattan Project Sites.

ERNEST MAZZAFERRI, M.D. (see above).

KATHRYN MERRIAM, PH.D., is the owner and president of two companies. The first, Synthesis, Inc., offers career counseling, personality style workshops, focus groups, ethnographic research into human behavior, and grant writing. The second is Project Turnaround, a non-profit corporation dedicated to working with high school dropout and at-risk youth. She is a past member of the local school district Board of Trustees. She has developed and implemented school curricula for the education of gifted students in California, Idaho, Connecticut, and Alaska. She is currently chair of the Bannock County Planning and Development Council, vice chair of the Bannock Planning Organization Citizen's Advisory Committee, Secretary of the Idaho Planning Association, and is on both the local and the state Board of Directors for the League of Women Voters of Idaho. She has received numerous grants to increase public involvement in issues related to nuclear waste, land use, and the deregulation of electricity.

DADE MOELLER, PH.D., is President, Dade Moeller & Associates, Inc., New Bern, NC. He earned B.S. (civil engineering) and M.S. (environmental engineering) degrees at the Georgia Institute of Technology and a Ph.D. (nuclear engineering) degree from North Carolina State University. From 1948 to 1966 he was a commissioned officer in the U.S. Public Health Service. During this time, he served as a research engineer at the Los Alamos and Oak Ridge National Laboratories; as Director, Radiological Health Training, Taft Sanitary Engineering Center, Cincinnati, OH; and as Director, Northeastern Radiological Health Laboratory, Winchester, MA. From 1966 to 1993, he was a member of the faculty at the Harvard School of Public Health. This included an initial appointment as Chair, Department of Environmental Health Sciences, and later as Associate Dean for Continuing Education. From 1973 to 1988 he served as a member of the Advisory Committee on Reactor Safeguards, U.S. Nuclear Regulatory Commission, and from 1988 to 1993 he chaired that Agency's Advisory Committee on Nuclear Waste. He is past-president of the Health Physics Society, a member of the National Academy of Engineering, an honorary member of the National Council on Radiation Protection and Measurements, and the recipient of the Meritorious Achievement Award, U.S. Nuclear Regulatory Commission, and the Distinguished

Achievement Award, Health Physics Society. Dr. Moeller is a registered professional engineer, he is certified by the American Board of Health Physics, and he is a diplomate of the American Academy of Environmental Engineers.

CHRISTOPHER NELSON, B.S., is an Environmental Engineer at the Environmental Protection Agency's Office of Radiation and Indoor Air. Mr. Nelson has worked for the EPA for over twenty-five years. His specialties are radiation risk assessment and air dispersion modeling. Mr. Nelson has authored or co-authored several published papers and, for his work with the EPA, has received four EPA Bronze Medals. Mr. Nelson is currently a member of the NRC's Committee on an Assessment of CDC Radiation Studies.

HENRY ROYAL, M.D. (see above).

RICHARD H. SCHULTZ, M.S., is the Administrator of the Idaho State Division of Health, Department of Health and Welfare. Mr. Schultz has worked in public health for over nineteen years, the last eleven of which he has served as the State Health Official for Idaho. Mr. Schultz has represented Idaho's public health interests in addressing radiation exposure issues associated with Hanford, Washington, and Idaho National Environmental Engineering Laboratory. He serves as a member of the Tri-State Executive Committee for the Hanford Health Information Network and as a member of the Association of State and Terrritorial Health Officials, Health Information and Core Public Health Policy Committee.

DANIEL STRAM, PH.D., is an Associate Professor in the Department of Preventative Medicine at the University of Southern California. Dr. Stram earned his Ph.D. in Statistics from Temple University, and subsequently engaged in postdoctoral research in Biostatistics at the Harvard School of Public Health. From 1986 to 1989 he was a member of the Statistics Department of the Radiation Effects Research Foundation in Hiroshima. Since 1990, Dr. Stram has been a major participant in NIH funded clinical research and epidemiology in childhood and adult cancers at the University of Southern California and the Children's Cancer Group. His radiation-related work in Hiroshima and U.S.C. has concentrated on statistical aspects of the dosimetry systems used for the A-bomb survivors and for the U.S. uranium miner's cohort study. Dr. Stram is presently a member of the NRC Board on Radiation Effects Research.

ROBERT G. THOMAS, PH.D., formerly of the Los Alamos National Laboratory, is a private consultant involved in lectures and workshops concerning the decommissioning, decontamination, and restoration of nuclear facilities. He attended the University of Rochester on a fellowship in radiological physics and subsequently received his Ph.D. in Radiobiology and Biophysics. Dr. Thomas was one of the planners and implementers in establishing the Inhalation Toxicology Research Institute in Albuquerque. He was an Assistant Professor at the University of Rochester and an Adjunct Professor at the University of New Mexico. His research interests focused on establishing acceptable guidelines for exposure to radionuclides. He led a team of radiological health experts into Romania, Russia, and the Ukraine immediately following the Chernobyl accident. Dr. Thomas is currently on committees for the National Council for Radiation Protection and Measurements and for the International Commission on Radiological Protection.